W0016985

U.S. ARMY FORD M8 AND M20 ARMORED CARS

CASEMATE | ILLUSTRATED | SPECIAL

CASEMATE | ILLUSTRATED | SPECIAL

U.S. ARMY FORD M8 AND M20 ARMORED CARS

DIDIER ANDRES

Acknowledgements

Many thanks for their help and support particularly to Alexandre Delbes, Michel Roques, Philippe Charbonnier and all the Histoire & Collections team, as well as Benoît van Pottelsberghe and Vincent Dumont de Chassart for access to the restoration of their M20 Armored Utility Car. Many thanks to TMWWII PDF Military Manuals for availability and access to its documents (http://www.tm-ww2.com/).

I especially dedicate this book to my wife Michèle for her patience over the past 20 years and to my most ardent supporters, Gabriel, Anaëlle, Bruno, Samuel, Clarisse, and Damien; may all of them be thanked greatly,
"The show must go on."

CISS0019

Published in the United States of America and Great Britain in 2023 by
CASEMATE PUBLISHERS
1950 Lawrence Road, Havertown, PA 19083, USA
and
The Old Music Hall, 106–108 Cowley Road, Oxford OX4 1JE, UK

This book is published in cooperation with, and under license from, Sophia Histoire & Collections. Originally published in French as *Les Automitrailleuses Ford M8 & M20 De L'U.S. Army* by Didier Andres © Histoire & Collections 2022

English edition © 2023 Casemate Publishers

Hardback Edition: ISBN 978-1-63624-310-8
Digital Edition: ISBN 978-1-63624-311-5

A CIP record for this book is available from the British Library

All rights reserved. No part of this book may be reproduced or transmitted in any form or by any means, electronic or mechanical including photocopying, recording or by any information storage and retrieval system, without permission from the publisher in writing.

Design by Myriam Bell
Translated by Alan McKay

Printed and bound in the Czech Republic by FINIDR s.r.o.

For a complete list of Casemate titles, please contact:
CASEMATE PUBLISHERS (US)
Telephone (610) 853-9131
Fax (610) 853-9146
Email: casemate@casematepublishers.com
www.casematepublishers.com

CASEMATE PUBLISHERS (UK)
Telephone (0)1226 734350
Email: casemate-uk@casematepublishers.co.uk
www.casematepublishers.co.uk

Contents

The Genesis of Light Armored Cars

1914–18

In the U.S. Army, the armored car was defined as a fairly light, wheeled reconnaissance vehicle, with a minimum of armor to withstand small-arms fire.

It had to be fast on the road and more mobile off road than normal wheeled vehicles, with enough armament to successfully engage unfortified enemy positions.

World War I

The first vehicles fitted with handmade armor appeared in 1898–1899; they could almost be qualified as the distant ancestors of the armored

Above: Taken at El Paso, Texas, this rear view shot of Armored Car No. 1 shows details of the compartments and how the armored bodywork was assembled. The two turrets were hand-operated and offset to remove blind spots as much as possible. (*U.S. Army*)

Right: Armored Car No. 1, built by the Thomas B. Jeffery Company in 1915, can rightly be considered the first vehicle of this type in the USA. This early example can be identified by its lantern "headlights," whereas the second vehicle's lights were housed within an armored fairing. (*U.S. Army*)

vehicle. They had their baptism of fire during the first months of 1914, in a period when a war of movement was still envisaged. As the conflict got bogged down in trench warfare, they were no longer relevant, so the armored car, as a concept, fell into neglect. Entering the war late in 1917, the USA was not very interested in this type of vehicle.

There were only five or six more or less successful experiments. At the time, there was a host of partly or totally armored car models. The name could not be settled on either: Scout Car, Armored Reconnaissance Car, Armored Car, Armored Motor Car, and a lot of others. However, only the vehicles built on a truck chassis could really be termed an "Armored Car."

Armored Car No. 2 was designed by the White Motor Company; its silhouette was more elegant and less imposing than the Jeffery model. To take its armor's considerable weight and get the most out of its 4×2 driveline, the rear wheels were doubled, which also ensured better off-road maneuverability. This early 1915 configuration shows this feature while also revealing little was done to streamline its silhouette! (*U.S. Army*)

317   ARMOURED MOTOR CAR   R. RUNYON

Armored Car No. 1 was made by the Jeffery Motor Company in 1915, on a chassis made by Quad. Its 4×4 driveline, with all-wheel steering, was driven from a main cab at the front of the vehicle and a secondary cab at the rear of the chassis.

The first example produced was used by the New York National Guard; a second was employed by the Army in 1916 during the conflict on the

Left: The 1917 version of Armored Car No. 2, with the superstructure made by the Van Dorn Iron Works this time, but still on its White chassis, presents a somewhat refined silhouette. The fenders have been enlarged, engine compartment ventilation improved, and the wheels protected by steel disks. (*U.S. Army*)

Above: America was keen on these new machines and, during Fourth of July celebrations, the public was able to approach and touch these precious vehicles; this is Armored Car No. 2. (*U.S. Army*)

Left: The Mack Armored Car was designed like a large boat, halfway between a scout car with its open-top fighting compartment and an armored car with its shields (here, removable) for machine guns. This late version was given a front panel with horizontal gills. The earlier panel was a solid piece of steel fixed above the large engine compartment ventilation opening. (*U.S. Army*)

Above: Assembled for a parade are two examples of the Mack Armored Car—in the foreground is the Locomobile chassis while the other is the White chassis. The main visible differences are the tires and front wheel arches, weaponry, the radiator protection, and the setback front axle of the White chassis. (*U.S. Army*)

Middle: The shape of the King Armored Car was more like the idea of what an armored car could be. Its silhouette was elegant, it had lateral protection bars against trees and bushes, and its spare wheels served as rollers to help prevent the vehicle becoming caught on an obstacle halfway along its body. (*U.S. Army*)

Below: The King Armored Car was the only vehicle of this type to take part in any foreign operations. Transferred from the Army to the United States Marine Corps, it was sent to the Caribbean when the Dominican Republic was occupied. (*U.S. Army*)

Mexican border. Armored Car No. 2, built at the same time, came from the White Motor Company. It was a lightly armored (0.2-inch steel thickness) 4×2 vehicle.

Three examples of the Mack Armored Car were built. The first used Mack's two-ton chassis, while White and Locomobile products were used for the other two.

Two other vehicles appearing in 1917 could be put in the same category, the King Armored Car, built by the Armored Motor Car Company,

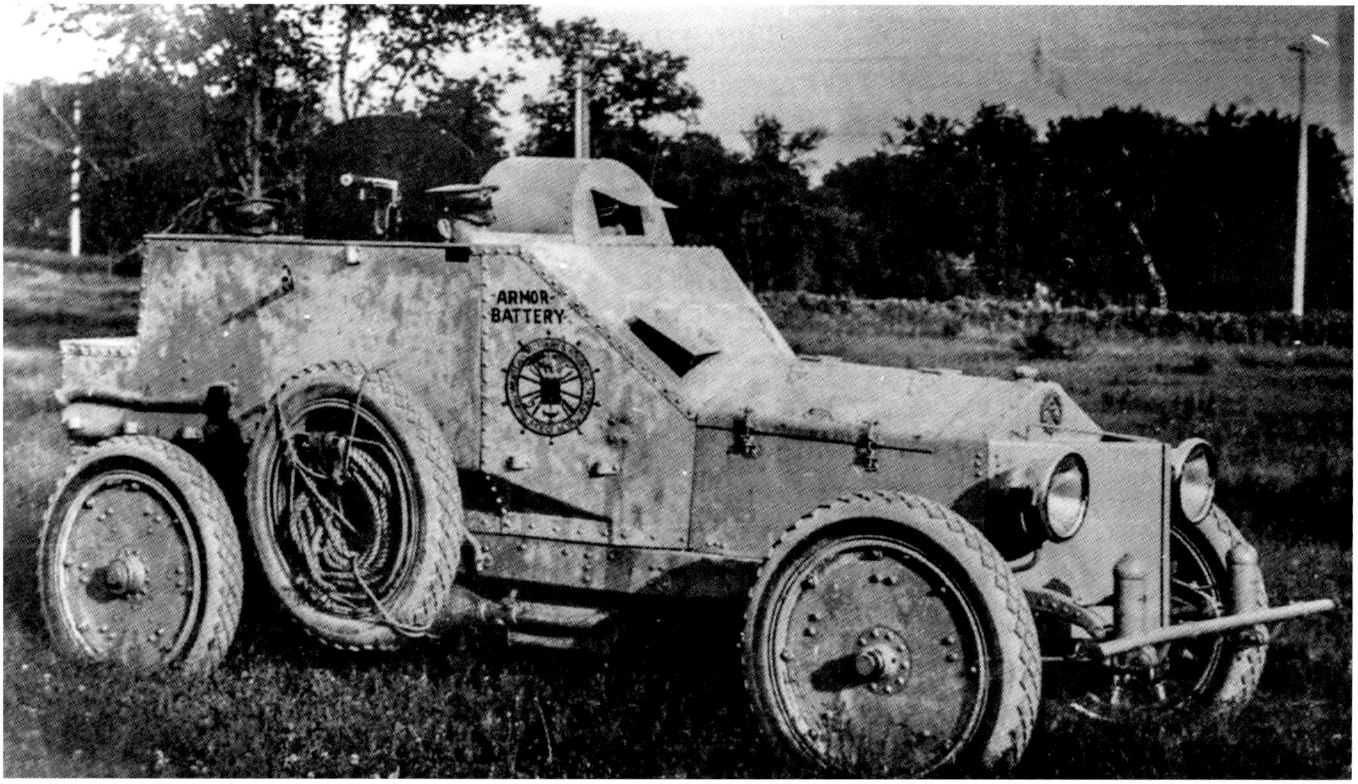
ARMOR
BATTERY

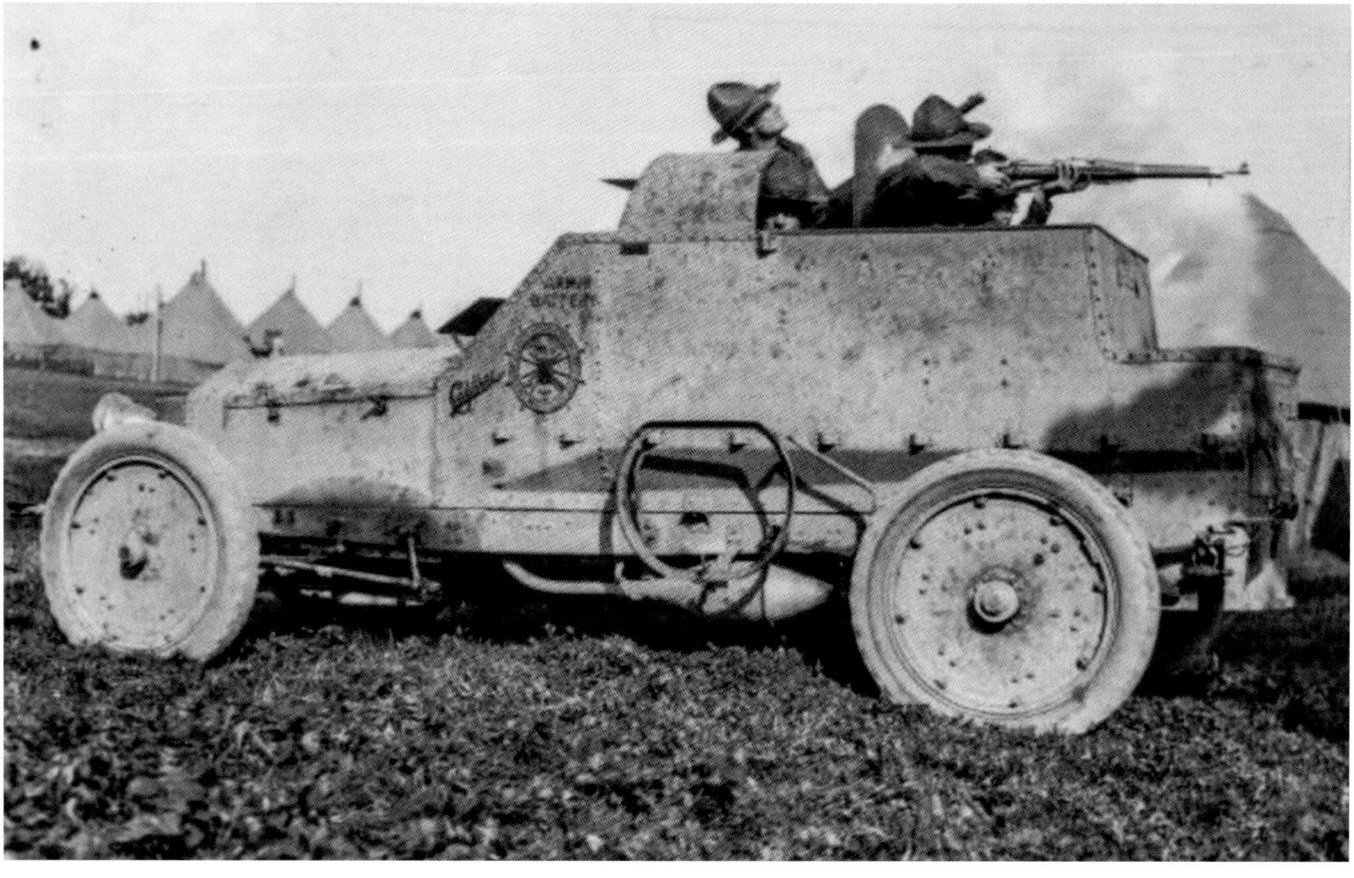

and the White Armored Car, which was but an extrapolation of Armored Car No. 2 with revised mechanicals and new armor. Other firms also had a go at the venture, Reo and Federal for example, but their performance was not as good. None of the vehicles really entered production, nor were any sent to Europe. At the end of the Great War, all these projects were suspended.

TECHNICAL SPECIFICATIONS OF THE FIRST ARMORED CARS

	Jeffery 1915	White 1915	White 1917	Mack	King 1917	Cadillac 1914
Length	216 ins	172 ins	—	236 ins	154 ins	—
Width	76 ins	64 ins	—	100 ins	81 ins	—
Height	96 ins	89 ins	—	76 ins	108 ins	—
Wheelbase	124 ins	130 ins	—	144 ins	120 ins	—
Weight	11,800 lbs	9,000 lbs	7,430 lbs	9,052 lbs	4,600 lbs	—
Load	800 lbs	—	—	—	680 lbs	—
Drive	4×4	4×2	4×2	4×2	4×2	4×2
Speed	20 mph	40 mph	45 mph	30 mph	45 mph	70 mph
Range	150 miles	—	—	—	200 miles	—
Engine	Buda	White	White	—	King	—
Power	4 cyl – 29hp	4 cyl – 36hp	4 cyl – 45hp	4 cyl – 45hp	8 cyl – 70hp 4 cyl – 50 hp	
Crew	4–5	4	3	5–7	2–3	4

Opposite: The only real pre-war realization was this vehicle using a Cadillac chassis. The study started in 1910 and went through a lot of modifications over the years, ending up in 1914 as this very early version, which finally inspired the King Company. (*U.S. Army*)

1928–34

The U.S. Army did not show any real interest in these armored vehicles for almost a decade. It was quite happy just to keep an eye on what was happening in other countries, remaining defiantly attached to its traditional cavalry units.

The first attempts were only made—very timidly—in 1928. Using a Pontiac chassis without most of its bodywork as a base, the Ordnance Department, working with Ford and Rock Island Arsenal, created two models of an experimental vehicle given the designation T1, one called "Scout Car" and the other "Light Armored Car."

The only armor was a shuttered guard in front of the radiator and a vertical plate in place of the

Above: The T1s, both scout and armored cars, were civilian vehicles without most of their bodywork. The base was a Pontiac while the mechanicals came from Ford, with a three-speed gearbox (and one reverse) and a 221-ci engine rated at 83 bhp at 3,800 rpm. (*Ordnance A.P.G. ref. 29097*)

Left: The T1 was simplistic; the only protection for the crew was a quarter-inch thick plate which partly replaced the glass windshield. (*Ordnance A.P.G. ref. 25318*)

windshield. Armament consisted of a .30-caliber machine gun mounted in front of the co-driver. The T2 Armored Car, which was the first vehicle in this category to be given comprehensive armor, was built on a La Salle chassis and four more variants were developed. The T3 designation was mistakenly given to the T1 Scout Car.

The 1932 T4 Armored Car (or Car, Armored, T4 in military parlance) was the first vehicle in this category to be standardized and mass-produced even though only 20 examples were made and required further development. The T5 prototype was a designation and vehicle transferred to a parallel category—Combat Cars with convertible drive (wheels and/or tracks).

The designations T6, T7, T8 and T9 concerned four designs by the Holabird Quartermaster

Above: The first version of the T2 looked more like a money transfer truck with side loopholes and an opening roof. With a 328-ci engine rated at 86 bhp, it reached a maximum speed of 70 mph thanks to its relatively light weight (4,854.58 lbs). (*Ordnance*)

Left: The T2E1's armor was more streamlined, with a .30-caliber machine gun mounted in a light turret swiveling through 360°. With its 0.13-in armor, it weighed 6,005.39 lbs. The T2E2's driving position was higher while the T2E3's silhouette was lower. (*Ordnance*)

Left: The James Cunningham, Son and Company built two examples of the T4 Armored Car before they were standardized. It was powered by a V8, 479-ci engine, rated at 133 bhp at 2,800 rpm, with a maximum speed of 55.30 mph because of its relatively light armor and 10,009-lb weight. (*Ordnance ref. 28034*)

Created by the Rock Island Arsenal in 1931, this prototype, originally designated T5 Armored Car, was re-classified as the T5 Convertible Armored Car before it became the T2 Combat Car. It could move around on its wheels, but fitting tracks gave it very good off-road mobility. With the seven-cylinder Continental engine it required for its 17,015.25 lbs, it had 6×4 drive. (*Ordnance ref. 38775*)

Depot, designed internally between 1929 and 1932. During this period of budgetary restrictions, the Army could not buy these vehicles from civilian companies, but there was nothing to stop it buying spare parts. Twisting the regulations slightly, four truck chassis and various disparate spare parts were purchased and then assembled to build these

four armored cars. The T10 was built on a Willys Overland Whippet chassis.

A single example of the T11 was built, from which six T11E1 models evolved; five were given to the Mechanized Cavalry for testing.

They could have been the second, standardized vehicles after the T4 but the design was too makeshift

The T6 Armored Car was one of the four vehicles of this type built by the Holabird Quartermaster Depot. This 4×4 vehicle was powered by a 95-bhp, six-cylinder Franklin engine, driving the vehicle's 7,206.91-lb mass to 60 mph. It was operational in 1932. (*Ordnance ref. 74927*)

The T7 had the same general features as the T6, differences being in the turret and the armament, a .50-caliber machine gun instead of the usual .30-caliber. (*Ordnance*)

and the mechanicals not reliable enough. There was no T12 Armored Car; the project remained on the Ordnance Department's drawing boards.

At this stage, the U.S. Light Armored Car project was interrupted again. There were again just two exceptions during this period when the country was neutral: the government authorized a vehicle to be built by American LaFrance in 1933 and another by Marmon-Herrington in 1934, to be delivered to the Persian government, but the customer approved of neither and the projects were abandoned.

The T8 was a simple 4×2 also produced by the Quartermaster workshops and tested in 1930. Entirely based on a Chevrolet chassis and engine, it had a crew of just two. Its 207-ci engine could move its 3,805.18 lbs at almost 55.30 mph. (*Ordnance*)

Above: The fourth and final Quartermaster project, the T9 Armored Car, was mounted on a Plymouth base. Its small engine (four cylinders) was only rated at 21 bhp and the vehicle was very slow and not very agile. (*Ordnance*)

Left: The T10 was an extrapolation of the T8 on a Willys Overland chassis and was not really an improvement on the Light Armored Car project. (*Ordnance ref. 76213*)

Below: The T11 was made by the Four-Wheel Drive Auto Company. Delivered to the Army in 1933, it was powered by a 353-ci Cadillac engine, rated at 115 bhp at 3,400 rpm, capable of moving its 11,261.21-lb mass at a maximum speed of 69 mph. The developments, T11E1 and E2, were not much different, even though six examples were built. (*Ordnance ref. 32681*)

Above: The example built for Persia by American LaFrance in 1933 had a 753-ci V-12 engine rated at 240 bhp at 2,800 rpm, which gave the 20,860-lb vehicle a top speed of 52.82 mph. Its mechanicals and bodywork did not withstand testing at the Aberdeen Proving Ground, Maryland, the test center for Army vehicles. (*Ordnance ref. 30566*)

Below: The second Persian armored car was designed by Marmon-Herrington in 1934. It was powered by a Hercules RXC 529-ci engine, rated at 115 bhp at 2,200 rpm, moving its 17,606.12 lbs at up to 44.74 mph. (*Ordnance ref. 31399*)

Above and left: There was no denying the T4's civilian origins given the chrome bumper and strange headlights. (*U.S. Army Signal Corps*)

Left: The Cavalry Board was very impressed by this 6×4's performance as it was far better than any other vehicle in its category, including the 4×4s. (*U.S. Army Signal Corps*)

The M1, the First Standardized Light Armored Car

The T4 Armored Car was built by the James Cunningham, Son and Company of Rochester, New York, in 1931. It was designed using the chassis and mechanics of a 6×4 civilian truck; two prototypes were built and tested at the Aberdeen Proving Ground the same year.

Below: New York, 1938, this M1 is being presented to the public. The girder used as a bumper is one of the things that differentiated it from the T4. (*U.S. Army SC-114552*)

They were generally satisfactory, but several minor changes were recommended so the vehicle could be taken on as the M1 Armored Car (Car, Armored, M1, in military parlance).

One can question, however, what prompted designing and pursuing a program for wheeled armored vehicles since, in this post-war period, the Army, and especially those responsible within the all-powerful infantry, did not see the point of them, except for short-distance reconnaissance. The Cavalry had to insist the Army continue the program. It hoped to equip itself with fast armored vehicles capable of carrying out long-range reconnaissance missions and providing mobile supporting fire for its advancing units. Evaluation progressed well with the two prototypes until 1934. However, in 1932, the Rock Island Arsenal began

Right: Aberdeen Proving Ground and its very challenging obstacle courses. Fitted with chains on the rear wheels, the M1 easily climbs this 45° slope. (*U.S. Army ORD-6381*)

a production series under the new designation "Armored Car M1." Apart from a few modifications to the rear running gear, reinforcing the shock absorbers, and positioning the large side storage lockers, the vehicle did not undergo any major modifications.

The body of the vehicle was made of armor plates whose thickness varied from one quarter to 0.375 of an inch, the assembly being welded and riveted. This gave the vehicle an empty weight of 9,835 lbs and a combat weight of 10,233 lbs. It was 180 inches long, 72 wide, and 83 high, with a 141-inch wheelbase.

The M1 Armored Car was powered by a 480-ci V8 Cunningham engine rated at 133 bhp at 2,800 rpm.

With its four speed (and one reverse) gearbox and 6.50×20 tires, the M1 reached a speed of 55 mph, its 30-gallon tank giving it a range of some 250 miles at a consumption of 8.33 mpg.

Driven by a four-man team, the M1 was armed with two machine guns, one .50-caliber and a .30-caliber, installed in parallel in a manually operated turret traversing through 360°, plus a

Left: Maintenance and inspection during 1934 maneuvers. This rear view shows how to get into the combat compartment and how the crew's equipment was stowed. The new storage lockers were positioned over the rear wheel arches. (*U.S. Army Signal Corps*)

Above: Between 19 April and 16 June 1934, the 1st Cavalry Regiment took part in important maneuvers between Fort Knox and Fort Riley. On 6 June, at the camp site, the vehicles were brought together to be inspected. No fewer than 10 M1 Armored Cars are present in this shot. (*U.S. Army Signal Corps*)

Above: A sad end for this M1 as it is used as a target on a firing range. (*U.S. Army Signal Corps*)

Left: Radio sets appeared very early in the cavalry units, even before the arrival of mechanical horsepower. Troop A of the 1st Cavalry Regiment. (*U.S. Army Signal Corps*)

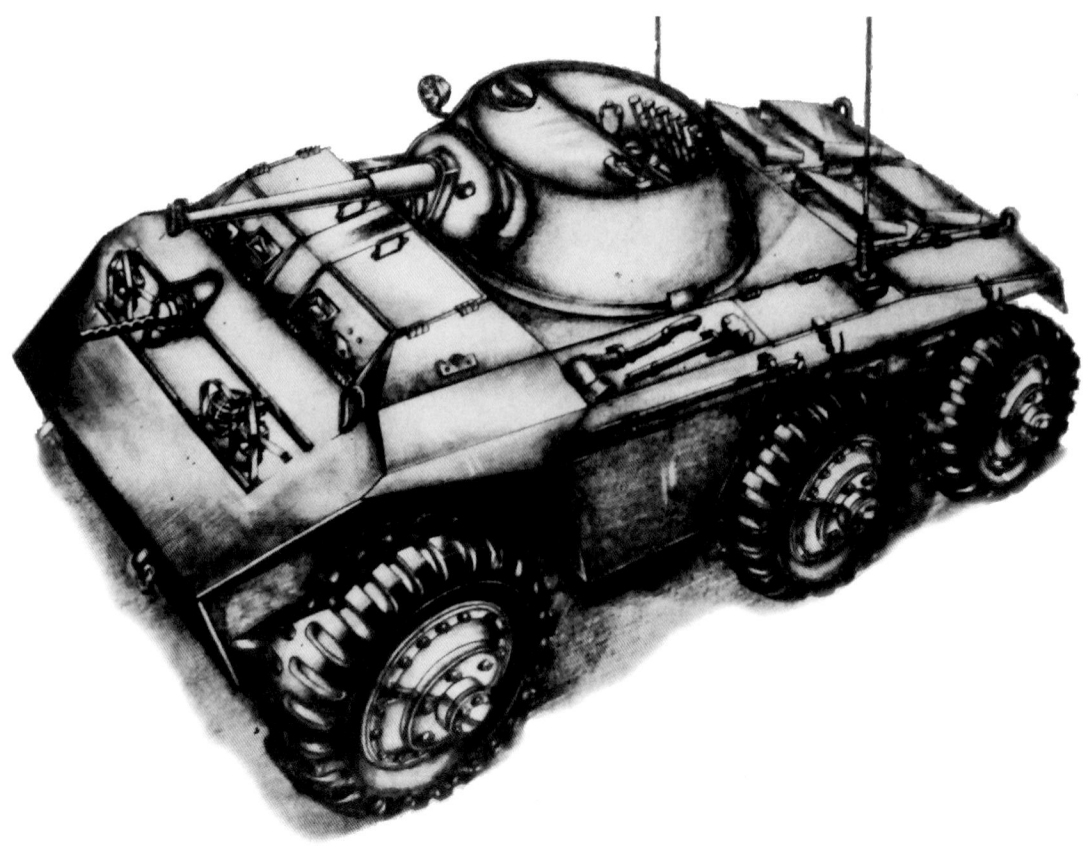

Right and below: The Ford technical offices presented these two artists' impressions of the future 37-mm Gun Motor Carriages: the 6×6 T22 and the T22E1 4×4 version. Both vehicles' basic silhouette was already quite distinctive. (*Ford Co.*)

.45-caliber machine gun for close-quarters defense. Another change from the T4 to the M1 increased firepower by installing a cradle on the left side of the turret to take another .30-caliber machine gun, which operated independent of the turret.

To ensure the vehicle's off-road ability and to help prevent it from getting stuck on an obstacle between the front and rear axles, an idler wheel was fitted on either side of the chassis. These also served as spare wheels.

Between 1931 and 1938, some 22 T4/M1 vehicles were produced, including the two prototypes. At first, they were assigned to the 1st Cavalry Regiment and took part in a lot of maneuvers whereas the two T4 prototypes did the rounds of the military academies, like West Point.

Subsequently, the M1s were shared among various infantry units which hung on to them until 1941, when they became targets on firing and training ranges.

The Fargo Division's (Chrysler) T23 prototype in its most advanced state of development, dated 15 May 1942 in the Detroit Ordnance District. (*Ordnance ref. A-T720*)

The driver's cab was covered with mobile armored hatches that did not impede the 37-mm gun when they were raised. (*Ordnance ref. A-T719*)

The T23E1 prototype from Chrysler included the same design features as the 6×6 version. It was only the number of wheels and the mechanicals that changed. (*Ordnance ref. A-T1011*)

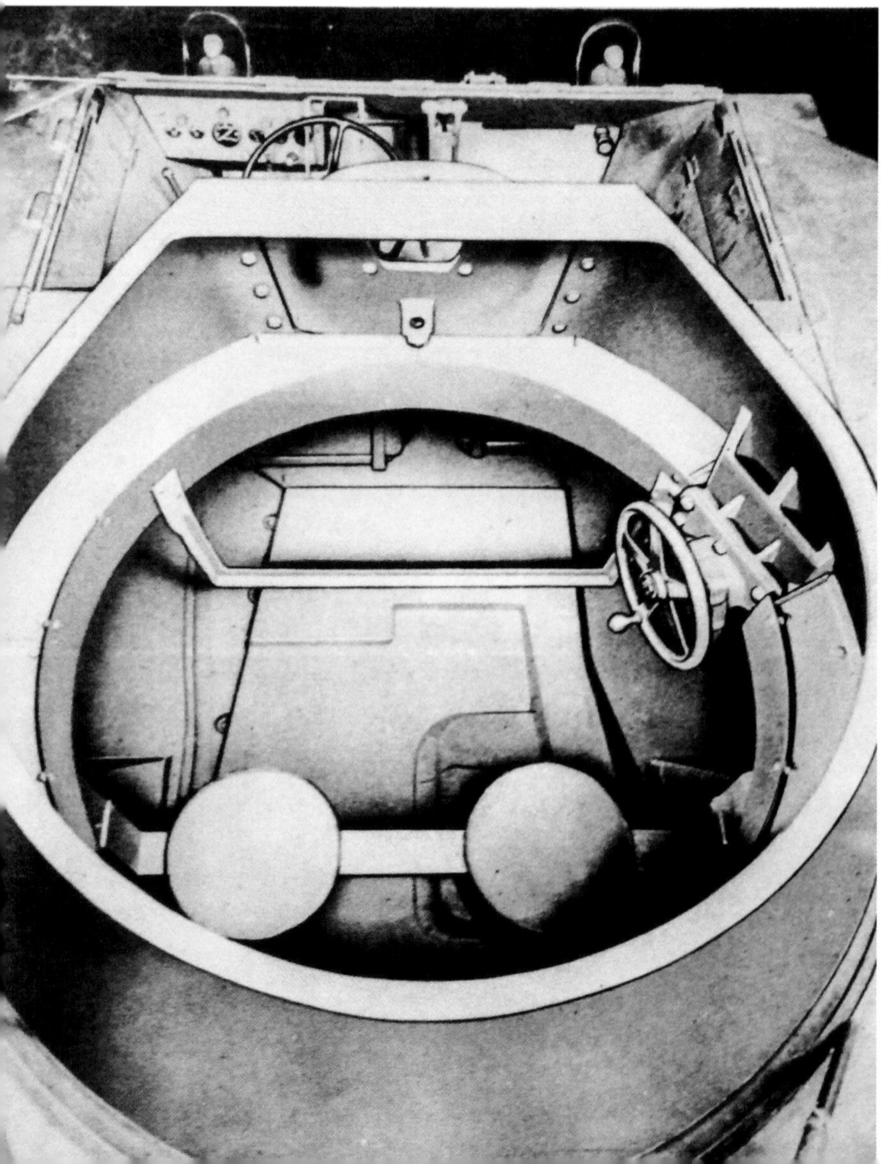

The Initial Projects, 1941–43

It was the evolution of the European war which saw a major re-think of all the United States' armament programs. In 1941, the first studies were made to equip the Army with wheeled vehicles armed with anti-tank guns.

Various leads were followed, all of them for creating Gun Motor Carriages capable of taking the 37-mm anti-tank gun. Two guidelines were envisaged: with and without armor. One of the projects in the second category was favored by the Army, resulting in the Dodge WC-55 Gun Motor Carriage. Two sub-categories appeared for the wheeled vehicles with armor: the first included the Medium and Heavy Gun Motor Carriages, among which, above all, was the Chevrolet Gun Motor Carriage T17E1 series.

Above: The T23 and the T23E1 were never armed and were tested as such. (*Ordnance ref. A-T1012*)

Left: Both the T23 and the T23E1 vehicles were identical as far as the turret and driver's position were concerned. (*Ordnance ref. A-T723*)

THE T23 DRIVER'S POSITION

1. Opening/closing of the engine compartment ventilation shutters
2. Handbrake
3. Instrument panel lighting
4. Hand throttle
5. Light switch
6. Clutch pedal
7. Unknown
8. Brake pedal
9. Accelerator pedal
10. Unknown
11. Reduction gear lever
12. 4×4 control lever
13. Gear lever
(*Ordnance Ref. A-T690*)

The second sub-category was for light vehicles, the Light Gun Motor Carriages, which became the Ford M8 and M20 described in this book.

The first studies started in July 1941. The technical features were described in a memorandum dated 30 July 1941, sent to the Chief of Ordnance on 5 August. The project talked about a Tank Destroyer Vehicle, a 37-mm Gun Motor Carriage, or even of an Armored Reconnaissance Car. Only on 12 March 1942 was this ambiguity solved by making the designation "Light Armored Car" official for the five projects being prepared at the time.

Basic Requirements

According to the specifications, the vehicle had to weigh about 10,000 lbs, be 176 inches long, 72 inches high and 82 inches wide, and be able to carry a crew of four. Its front and turret armor had to withstand .50-caliber machine-gun fire from 250 yards and its flanks had to be able to resist .30-caliber machine-gun bursts fired from 100 yards; protection of the other parts of the vehicle were left to the discretion of the manufacturer.

The main armament had to be the standard anti-tank 37-mm cannon, with a co-axial .30-caliber machine gun, both mounted on a single assembly fitted into a turret rotating through 360°, the vertical clearance for the whole assembly being -10° to +20°. The secondary armament had to include a .30-caliber machine gun in a ball and socket mounting near the driver's position and four .45 machine pistols for the crew.

The new vehicle's mechanical performance had to allow it to maintain a speed of 15 mph on a 10 percent slope, a speed of 55 mph on a tarmac road, and 35 mph off road. Its 12-inch ground clearance had to enable it to overcome a 12-inch vertical obstacle and wade through a 24-inch ford.

The vehicle was driven by the four rear wheels through a four-speed gearbox and two-speed transfer box, enabling it to go from 6×4 to 6×6, and back, with the throw of a lever. The interior had to have enough room for storage, racks for shells, and space for installing radios. The engineers had to design the vehicle so it could also be used for other tasks without modifying the basic design. It was thought desirable that other versions of the vehicle could be fitted with a double or quadruple .50-caliber machine-gun mount for ground support, a twin 20-mm anti-aircraft mount, an 81-mm mortar, or be adapted for carrying ammunition onto the battlefield.

Above: Aberdeen Proving Ground, 16 November 1942. The Studebaker T21 poses for its technical specifications file photo. (*Ordnance ref. 73684*)

TECHNICAL FEATURES OF THE PROTOTYPES

	Specifications	Studebaker T21	Ford T22	Fargo T23
Length	176 ins	202 ins	190 ins	200 ins
Width	82 ins	80 ins	91 ins	90 ins
Height	72 ins	74 ins	80 ins	79.75 ins
Wheelbase	—	—	128 ins	109 ins
Weight	10,000 lbs	17,200 lbs	15,000 lbs	14,500 lbs
Turret rotation	360°	360°	360°	360°
Gun clearance	-10° – +20°	-12° – +25°	-10° – +20°	-10° – +20°
Max. Speed	55 mph	55 mph	55 mph	48 mph
Range	—	280 miles	300 miles	Unknown
Drive	6×6	6×4	6×6	6×6
Powerplant	—	Hercules JXD	Hercules JXD	Chrysler
Power	—	112 bhp	112 bhp	105 bhp
Gears	—	5+1	4+1	5+1
Tires	—	Unknown	9.00 x 20	12.00 x 20
Electricity	—	24 volts	6 volts	12 volts
Crew	4	4	4	4
Vertical obstacles	1 ft	12 ins	12 ins	Unknown
Fording	2 ft	32 ins	24 ins	Unknown
Ground clearance	1 ft	Unknown	12 ins	Unknown

The Project's Evolution

With OCM No. 17303 dated 9 October 1941 from the Ordnance Committee and OCM No. 17359 dated 23 October 1941 from the adjutant-general, the design of the two 37-mm Motor Gun Carriage prototypes, designated T22 and T23, was entrusted to two different builders.

In November 1941, contracts were signed with the partners who were thought the most likely to meet the project's parameters. The Ford Motor Company received the order to develop the T22 and the Fargo Division of the Chrysler Corporation was to progress with the T23.

OCM No. 17515, dated 11 December 1941, authorized the purchase of two examples each of the T22 and T23 prototypes. For the sake of efficiency, it was also decided to extend the study by pitting six-wheel vehicles against four-wheel vehicles. Two examples of the latter arrangement, designated T22E1 and T23E1, were also ordered.

In March 1942, the first of the two T22s was considered ready and was authorized to go to the Aberdeen Proving Ground on the 16th for a series of demonstrations before going on to Fort Knox for testing. Meanwhile the Service of Supply recommended the model be adopted; the second T22 ordered was therefore abandoned.

At the beginning of April, in Note No. 451.24/2210, the Service of Supply, basing itself on the progress made on the T22 and the recommendations made by the Armored Force and Army Command, suggested the T22 project be ratified as soon as the requested modifications were carried out, thereby indicating the T23 program would be cancelled. Immediately afterwards, on 14 April 1942, in Note No. 451/4923, the cancellation of the T22E1 and T23E1 programs was confirmed.

The M8 Light Armored Car

The T22 Gun Motor Carriage was effectively adopted under the name of Car, Armored, Light, M8 in June 1942; this decision officially put an end to the T22E1, T23 and T23E1 projects.

The prototypes that were almost finished, however, could be completed. These had to be handed over to the Ordnance Department since it was considered unprofitable and unjustified to stop building them. Although Chrysler only built one T23 and one T23E1, history does not say what happened to them, but it can be assumed they went the same way as the Ford vehicles. On 24 October 1944, the Detroit Ordnance District ordered the sole T22 and the two T22E1s to be transferred from the American Car and Foundry Company, where they were stored, to the Ordnance Service Command Shop at Fort Custer for scrapping.

The Studebaker Wonder

Launching new projects always roused the industrialists' greed. Although eliminated from

Above: Compared with rival vehicles, the T21 had a complicated rear end with critical components exposed. (*Ordnance ref. 73684A*)

the 37-mm Gun Motor Carriage project, the Studebaker Company put forward a vehicle built to the required specifications at its own expense. The report by the special Armored Vehicle Board recognized that this prototype, initially designated as the T43 and then the T21, would be a satisfactory reconnaissance vehicle but there were too many design faults, and the factory was unable to promise it could get the design into production within the year. The Studebaker T21 was nonetheless delivered to the General Motors Proving Ground on 22 May 1942 to undergo tests. As its shape and intended role were too much like the T22, the project was abandoned.

Even with the help of the document "Design, Development, Engineering and Production Report of Armored Cars," it is not easy to discover why the Ford T22 was adopted instead of the T21 or the T23. For the Fargo T23, it is easy to note the failure to meet deadlines; that there were fears about its traditional chassis, whereas the T22 had a load-bearing monocoque body; that there was a problem with the drivers' positions interfering with the gun's ability to fully depress; and that the vehicle's lower top speed was due to it being underpowered.

For the Studebaker T21, opinions were more divided and less clear. Setting aside that this firm was not mentioned in the project and that its participation was against the regulations the Army had to follow when purchasing materiel, the T21 looked good and the prototype was visually complete. Its quality was unanimously recognized as excellent, especially with its suspension consisting of independent wheels and torsion bars. However, the mechanicals were thought to be too complex for its intended use. The Ford T22 was therefore the logical choice.

2

Car, Armored, Light, M8

The T22 Prototype

Right: March 1942, the prototype T22 Gun Motor Carriage was finished, ready to go through the experts' hands. Its general lines resemble the final M8 Light Armored Car produced by the Ford Motor Company. (*U.S. Army SC-130857*)

Below: The T22 Gun Motor Carriage at Fort Knox, Kentucky, 22 March 1942. The only markings are the three letters "USA" on its left rear fender. It had just finished its first off-road tests, during which it lost its muffler and the end of its exhaust pipe! (*Ordnance ref. 2728*)

In the end, a single T22 was thought suitable to go through the countless phases of validation. Taken on charge at the beginning of March 1942, it was transferred to the Aberdeen Proving Ground on 16 March to undergo its first in-depth examination.

On the 23rd of the same month, a panel of experts—consisting of members from the Armored Force, the Cavalry, Tank Destroyer Command, and the Ordnance Department—was given the task of examining the new vehicle in great detail. The result was a 28-point memorandum, dated the 25th, that listed all the remarks that would affect

the T22's development on its way to becoming the T22E2.

The following are some extracts of the memorandum:

1. The muffler was torn off during the cross-country operation and has to be repositioned and the exhaust pipe's path modified.
2. The main braking system has to be modified because it was not effective enough; using the new Bendix system is possible, replacing the master cylinder and adding a vacuum booster (recommended).
3. The handbrake mechanism is too exposed to mud thrown up against it and to shocks; it must be protected and completely enclosed if possible.
4. Installing a rev counter (tachometer) is necessary and, if possible, be the standard model used on tanks. An indication panel marking the maximum rpm, the maximum speeds, and the calibration of the speeds in relation to the gear chosen must be supplied.
5. The electrics have to be changed from 6 to 12 volts, a 180 amps/hour battery like those used on tanks has to be placed in the engine compartment. Failing that, two 6-volt, 120 amps/hour batteries can be used.
6. A separate generator will be essential because of the new radio equipment requirements. While

waiting for further information, a 750-watt Delco Motor Generator, like the one suggested for the M5 Light Tank, is to be used. The exhaust silencer should be independent and perform extremely well.

7. The fuel capacity should give it a range of at least 300 miles. The Armored Force Board would want 500 miles. The vehicle's present capacity should be increased by at least 100 miles, as long as there is available space for the tank.
8. The lower armor plate at the front of the body should be widened on the right and left to

Above: Before the T22 left the factory, this series of shots was taken to illustrate the technical file which had to accompany the vehicle. On this front left three-quarter view, the vehicle is in operational configuration with the drivers' hatches opened. (*Ford Co. ref. 76641-E*)

Below: Taken from the front with the hatches closed reveals the interference it caused to the gun barrel's ability to fully depress. Photos could be touched up at the time, as seen here with the headlight protection drawn on. (*Ford Co. ref. 76641-A*)

Above: This rear view allows one to see how badly positioned the exhaust muffler was. (*Ford Co. ref. 76641-D*)

protect the suspension, the shock absorbers, and the outer parts of the steering mechanism. The front wings that form the fender should be lengthened forwards and downwards to correspond with the extension of the new frontal armor plate.

9. Bulges must be fitted on each side of the vehicle between the front wheels and the leading wheel of the rear bogey. They must stick out from the present sides of the body by 15 inches. The present toolbox on either side of the body must be kept and will be situated under the bulge which takes the radios. They do not have to be made with armor plate, but they must be thick;

a half inch for the front, side and back faces, and a quarter inch for the lower and upper surfaces.

10. The following radios must be installed: in the right bulge, an SCR-506 or SCR-193 and, in the left one, an SCR-508 or SCR-245. Everything must be done so an SCR-293 can be placed in any of the locations described.

11. The radio aerial must be moved rearwards on another bracket on the right side of the body. Its position must be incorporated into the body, or above the rear tandem wheels, and the base and antenna mast protected.

12. Protection for the headlights and rear lights must be installed.

13. A 5-ton tow pintle must be fitted to the rear of the vehicle.

14. To make loading on rail wagons or ships easier, lifting eyes must be placed at each corner of the body.

15. Rear and front towing lugs must be positioned at the height and spread as prescribed for the Light Tank.

16. A drain plug is needed in the front part of the hull; it must be the standard type.

17. A study must be made to envisage reinforcing the front part of the floor forward of the driver's seat.

18. All technical openings in the floor and the hull must be sealed.

19. The order in which the gears are selected is not standard and must be modified. An indication H-plate with a diagram must be fitted.

Comparing the size of man and machine: a photo taken on 18 March 1942 at the Aberdeen Proving Ground, two days after the vehicle's arrival, shows the promise to build a low-profile vehicle has been kept. (*Ordnance ref. 59444*)

Above left: The internal layout of the T22 37-mm Gun Motor Carriage turret was special and completely different to the final version that equipped the M8s. The seat supports were improved and moved to the rear of the turret; the turret rotating wheel was moved to the left side and was therefore the responsibility of the gunner, not the car commander. (*Ford Co. ref. 76640-B*)

Above right: The transfer box was located under a crosspiece in front of the transmission. The rear engine design meant a linkage system was needed to shift gears, the gear lever being situated at the front. The 23 March meeting concluded that closing off and sealing this open space of the vehicle was imperative. (*Ford Co. ref. 76640-A*)

20. Using the clutch pedal is not satisfactory and its position prevents the driver from resting his foot on the floor.

21. The base of the turret ring must be marked off in 10° intervals and an index pointer installed.

22. Side protection against any objects being thrown up must be installed on all wheels and cover the sides down to the hub, while being careful not to hinder the clearance of the steerable wheels.

23. The front and rear wings have to be horizontal rather than sloping as they are now; storage boxes are to be installed above the twin rear wheels and, if possible, over the front wheels. They must open upwards and their covers must prevent water from entering.

24. Chains must be available for all six wheels.

25. The turning circle is not satisfactory and everything must be done to reduce this. The width of the front tread is 74 inches and the rear is 76 inches; they must be the same. Standard non-directional tires must be fitted as directional tires are no longer standard.

26. The possibility of fitting a system on the front wheels which would serve as a capstan. This recommendation will wait for the result of

the ongoing trials at Fort Knox, based on the amphibious 0.75-ton model by Chrysler, the Aqua Cheetah.

27. Evaluation of the Turret

(A) The M23 Gun Mount must be used; the turret rotation mechanism must turn from right to left, activated by the gunner's left hand. A new, smaller, rotation mechanism is desirable. It must be like that of the M3 Medium Tank and located in front of the gunner's left shoulder. The position of the riflescope must be analyzed; if satisfactory, it must be left as such. The roof of the turret must be enlarged to cover roughly half of its surface together with a periscope identical to the one used by the M5 Light Tank. A periscope rotating through 360° must be installed on the right of the turret, in

Below: All the driveline levers were positioned in a connecting box: a gearshift for the four forward and one reverse speeds, a lever for engaging the front axle, and another for the two speeds of the transfer box. (*Ford Co. ref. 76680*)

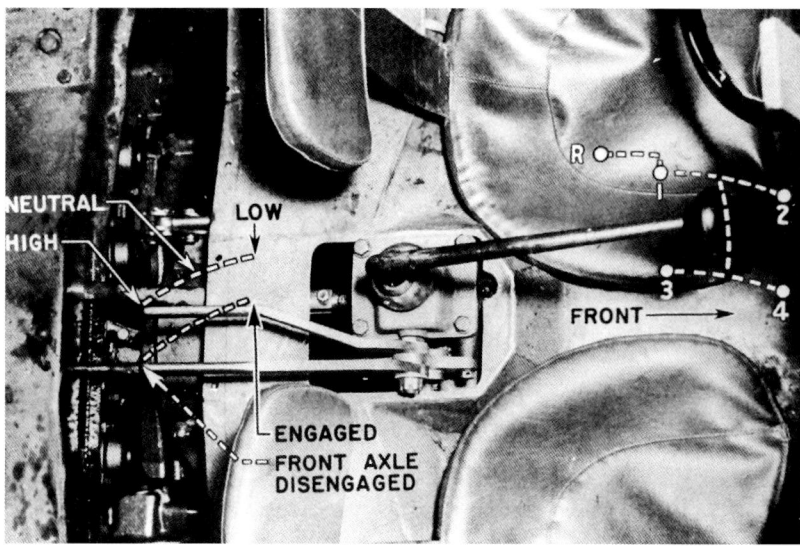

alignment with the vehicle commander/loader's sight. The height of the turret must be increased by 2 inches to enable the gun to recoil in its fully depressed position. An identical system to that of the M1 Combat Car could be installed; it would enable part of the roof to be opened forwards.

(B) Both turret seats must be installed on new supports moved to the rear and extended until their framework meets under the cannon. At this junction, a spent shell case box must be fitted and, a little further back, footrests for the two servers must be fitted. The height of the seats must be adjustable and lockable when they are in the high or combat position.

(C) An AA gun carriage for a .30-caliber machine gun, like the M3A1 and M5 Light Tanks, must be positioned on the right rear of the turret for use by the car commander.

28. Evaluating the Body

(A) The present driving compartment design is quite unsuitable: when closed, the casemate prevents the 37-mm gun from being lowered. The pivoting rear and side casemate plates must be rigid and welded in place. The height and width of the foldable plates on the front will be reduced and they must be separated by a fixed support connected to the rear plate. The roof plates will pivot outwards, each fitted with a lock. On the front plates, the driver's periscope must be situated slightly to the right, and the direct vision slots slightly to the left; the opposite layout is to be fitted for the co-driver.

(B) Armored stoppers for the fuel and water tanks must be installed.

(C) The air intake and outlet protections at the rear and on the top of the engine compartment are inadequate; they leave part of the radiator visible and do not protect against light weapons fire.

(D) Access to the radiator and the air filter is complicated; a solution must be found to improve this.

(E) The direct vision slots on the front of the driving compartment must be able to take transparent vision blocks and be covered by a removable armored plate.

(F) A study must be made on the .30-caliber machine gun and on the ability to maintain it (it was finally abandoned before the pre-series vehicle was built).

It was interesting to discover this analysis of the Ford prototype presented by the experts from varioU.S. Army commands. What is very surprising, however, is the imperative verb form used throughout, leaving no doubt as to what the military wanted. These were not wishes, but orders. With just a few comments, the vehicle was redesigned. What is also surprising, however, is there were no opinions given about the mechanics. The initial trials were therefore conclusive and the vehicle was presumably "well conceived."

The second last part of the memorandum repeats the desire to see the vehicle produced in four versions: 37-mm Gun Carriage, Command Car, Anti-Aircraft Car, and Personnel or Ammunition Carrier. The end of the report listed various points Ford had to explain, a parts list, and the documentation the Army was to make available. A final request was also made that Ford consider using a Continental or White powerplant; the choice of engine was ultimately the manufacturer's prerogative and the Army could only make suggestions.

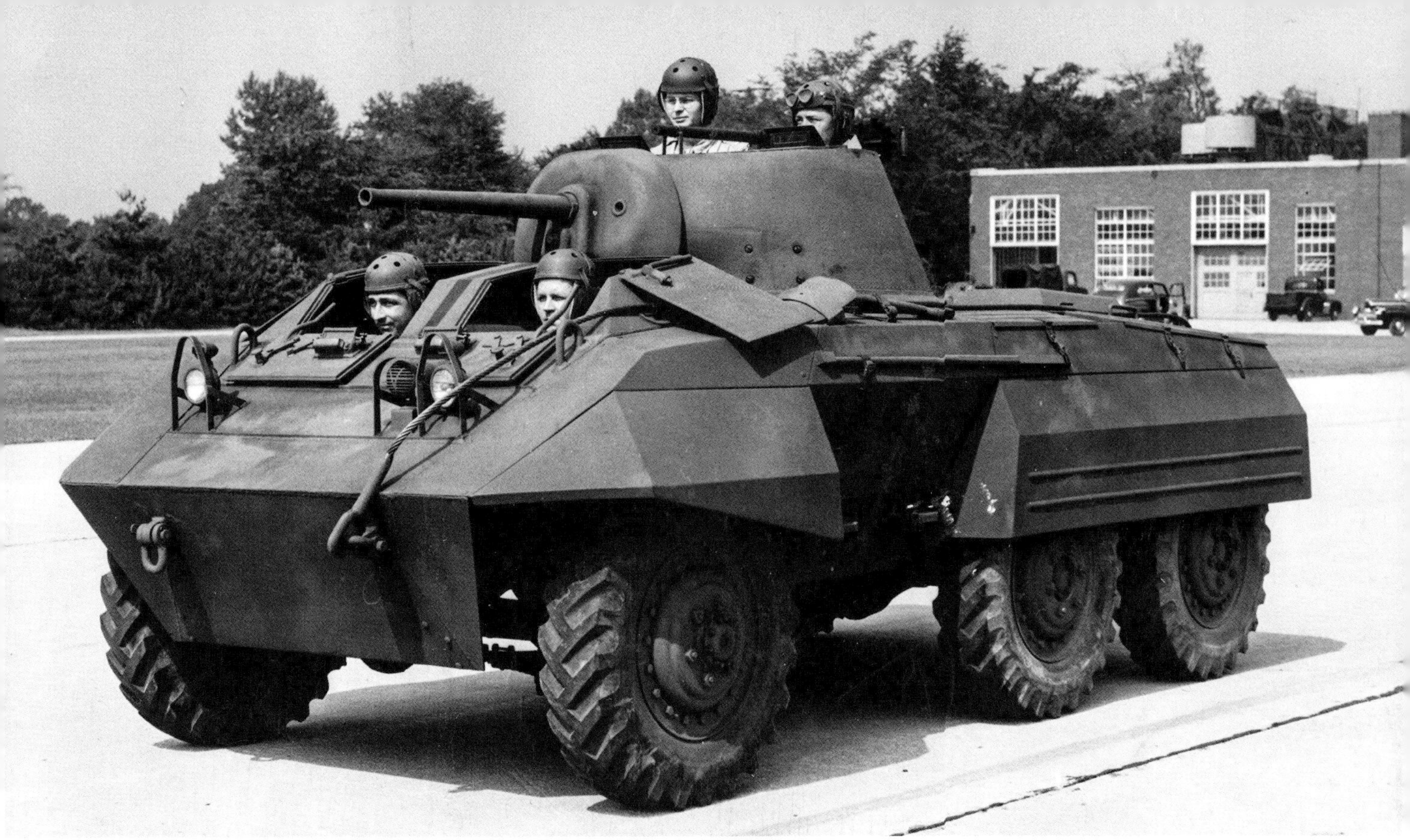

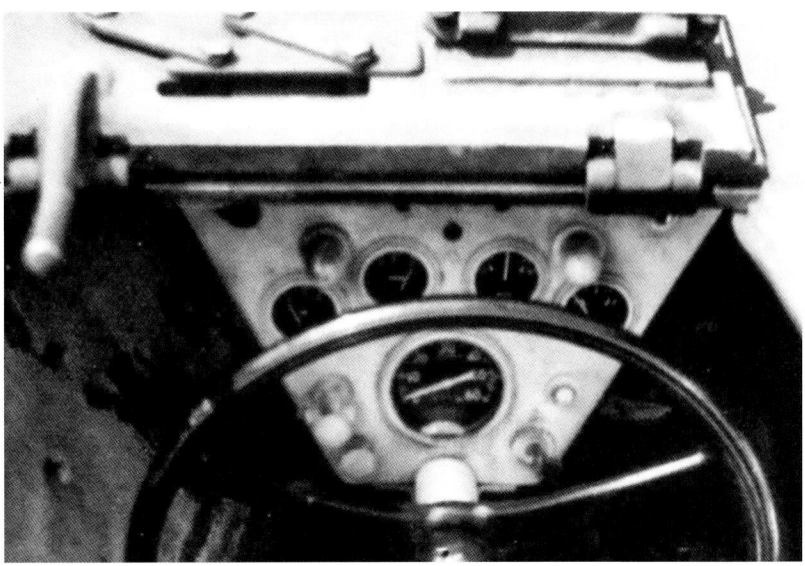

The T22E2 Pre-Series Vehicle

Above: The pre-series T22E2 out of the workshop for its first presentation, 29 June 1942. It is still not finished; its armament is fake as the Army had yet to supply the mount, the carriage or the 37-mm cannon. (*Ford Co. ref 76844*)

At the 23 March 1942 meeting with the Army experts and representatives from the factory, the Ford Motor Company was allowed to modify the vehicle that had been used to present their project in 1941.

The Army was in a hurry and wanted a delegate from the Ordnance Department to check on the 26th or 27th that the model had been realistically completed. If this was the case and progress considered satisfactory, another meeting with the Armored Force, the Cavalry, Tank Destroyer Command, and the Ordnance Department would be organized for the week of 30 March.

Making the pre-production series vehicle was left to the factory—it could either build a new vehicle or use the second T22 prototype. The Ordnance Department was conscious of the enormous

Right: The T22E2's instrument panel was special because of its shape and the way the dials were laid out, except for the rev counter required by the Army. (*Ford Co. ref 76844*)

Above: Several details distinguished this vehicle from the mass-produced model, such as the layout of the crew's tools or the protective cover on the fuel tank filler. The half-covered turret was also to be modified. (*U.S. Army Signal Corps*)

Above: The basic outline has been finalized, such as the driver's casemate, with its four armored hatches. The roof plates have not yet had their side rests fitted and the siren is still on the left next to the headlight. (*U.S. Army Signal Corps*)

Right: The first official portrait taken at the General Motors Proving Ground on 18 August 1942. (*U.S. Army Signal Corps*)

pressure and the high stakes involved. With such short deadlines, nobody was expecting the vehicle to be complete in all details; in fact, all that had to be done was incorporate the changes listed in the 25 March 1942 memorandum.

At the end of the month, and with the new model being approved, Ford did everything possible to build the T22E2, helped by representatives from the Ordnance Department who did not spare their efforts. They had the difficult task of coordinating the efforts between the factory and the Army to ensure the various parts Ford was expecting were indeed in the process of being delivered.

The first major step was made on 22 June with the new body, mounted on its running gear, finishing its company tests on the trials track and returning to the factory. Even though no accessories were installed and the turret was incomplete, the T22E2 was officially a working vehicle.

On Tuesday, 23 June, despite of a lot of reminders, the Army had still not supplied the M23 gun carriage or the M6 gun but promised it would be sent before the end of the week. In the factory, interior equipment was fitted and the stowage coffers for the periscopes and hand grenades were installed; the water jerry can fittings, the first-aid box, the fire extinguisher, and the maintenance gear for the cannon were in order. The other mountings and boxes were being designed and installed.

Between 24 and 30 June, all attention was directed towards getting the turret ready. It was installed, as was the swiveling ring. Fitting the M23 carriage was more complicated; the dimensions given were not correct and it was necessary to rework the seats to make it fit. The manual mechanism, the ammunition box, the shield and periscope were positioned. The 37-mm gun had still not been supplied.

Between 3 and 13 July, the vehicle underwent its first road tests with Ford and was then ready to be transferred to the General Motors Proving Ground.

Rolling Tests and Collecting Data

From 13 to 16 July, the T22E1 was at Ford Airport to undergo long rolling tests to make sure the cooling system was working correctly, the carburetor was properly tuned, the brakes worked, and to collect the first set of operating statistics.

Since the transfer order was not forthcoming, extra powerplant and braking tests were programmed internally and the first quantified data were obtained. When functioning normally between 35 and 40 mph, fuel consumption was between 7 and 8 mpg.

Nothing much happened between 24 July and the end of August, apart from tuning and attention to details and issues as they arose. The gun was finally delivered and installed. An initial inspection by a representative from Washington, a Lieutenant Hichert, was made, and there were a few remarks, but, generally, a feeling of enthusiasm predominated in his report.

The Road Trip to Fort Knox

On Monday, 31 August, at 2:30 pm, the T22E2— cleaned and entirely repainted—took to the road under its own steam for Fort Knox, under the responsibility of Mr. McDonald, representing the Ford Company. Five hours later, the crew halted in the small town of Piqua in Ohio to spend the night after a first stage of 180 miles. Since the Army had not given the vehicle a registration number, and since a military vehicle could not travel on public roads without an escort or a registration number, the factory hastily painted "USA-W-6038230"

Top and middle right: Monday, 31 August 1942, the T22E2 leaves the factory to be delivered to Fort Knox under its own power. A 180-mile drive got it to the small town of Piqua, Ohio. (*Ford Co.*)

Bottom right: At midday, during a roadside break. Mr. McDonald, a Ford engineer, poses in front of the vehicle before setting off again. (*Ford Co.*)

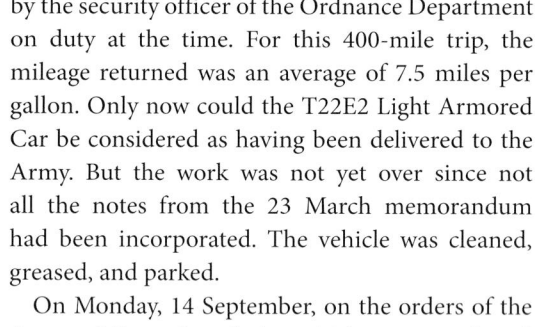

Above: The shape of the front fenders was the only major change to the outline; the long side triangle was reduced and no longer took up the whole length of the fender. (*Ford Co.*)

on the T22E2. The number was not chosen randomly, but from the numbers allocated to future production-series vehicles.

On Tuesday, 1 September, at 8:30 am, the crew set off again and traveled the remaining 220 miles to Fort Knox. The T22E2's first trip ended six hours later without any hitches. The rest of the day was spent with the vehicle being completely inspected

by the security officer of the Ordnance Department on duty at the time. For this 400-mile trip, the mileage returned was an average of 7.5 miles per gallon. Only now could the T22E2 Light Armored Car be considered as having been delivered to the Army. But the work was not yet over since not all the notes from the 23 March memorandum had been incorporated. The vehicle was cleaned, greased, and parked.

On Monday, 14 September, on the orders of the Armored Force Board, the vehicle was transferred to Camp Hood at Tempe, Texas. Strangely enough, the vehicle was delivered with a broken front axle transmission, without any explanation provided. On Wednesday, 13 October, the T22E2 was transferred to the Aberdeen Proving Ground to undergo tests and an inspection by Army Ground Forces.

On Friday, 6 November, Ford asked to be dispensed from installing the rev counter on the instrument panel, imposed in March. At the end of the year, 1942, the factory was already working on getting the assembly lines ready for series production and the design office likewise on the construction details. This request generated a favorable answer, received four days later.

Left: Tuesday, 1 September 1942, the T22E2 and its crew ready to set off on the remaining 220 miles of their trip from Piqua to Fort Knox. (*Ford Co.*)

Finalizing

November 1942 was a crucial month for developing what was now known as the M8 Light Armored Car. All the remarks and requests were changed into a programmed list of production changes. The T22E2 was now just a visual reference; the pre-series vehicle physically evolved no further.

A final series of tests was carried out to answer queries about the secondary armament. A .30-caliber machine gun was installed on the turret nape, its mount being directly inspired by those on the M3 and M5 Light Tanks and adapted to the turret slope angle. It was intended for the vehicle's close-quarter protection through 360°. For AA defense, a .50-caliber machine gun was mounted on a short column on the forward half of the turret, but trials showed it was almost impossible to use it beyond a very much reduced firing zone in front of the vehicle. Both arrangements were abandoned.

THE MACHINE-GUN TURRET MOUNTS

Top: In June 1942, the T22E2 37-mm Gun Motor Carriage became the M8 Light Armored Car. The vehicle poses here at the Aberdeen Proving Ground on 16 November 1942. (*Ordnance ref. 73681*)

Middle: This rear view shows the particularities of the T22E2 with its stamped rear fenders and the missing towing hook. The lack of hinges on the side skirts means they have to be dismantled completely to reach the wheels and mechanicals. (*Ordnance ref. 73682*)

Bottom right: This bird's eye view gives a good idea of how difficult it was to use the 50-caliber machine gun through 360° while remaining inside the turret. (*Ordnance ref. 73687*)

(1). The external mounting for the .30-caliber machine gun on the light tanks was a tube with two lugs bolted to the turret.

(2). On the T22E2, the mounting is identical but fitted to a spacer to account for the slope angle.

The First M8 Light Armored Cars

Two reports shaped the project's final silhouette. The T22E2 was delivered at the end of August 1942; four months had now passed and all parties concerned had easily had the time to examine the vehicle and test it in various conditions.

The Aberdeen Combat Vehicle Test Board handed in its conclusions on 27 November 1942 before the factory started getting ready for mass production. These remarks concerned the position of the gear lever, the slope of the steering column, improving the turret swiveling mechanism, installing an interphone between the combat and driving positions, installing drainage plugs under the body, the bolts for the turret swivel mechanism, and

Above: At the Highland Park Ford factory (Saint Paul, Michigan) in March 1943, the very first M8 Light Armored Cars came off the production line. The factory took advantage of the opportunity and parked the sixth vehicle built on its forecourt for the public, and especially its workers, to see. (*Ford Co. ref. 77476-24*)

Left: The design of the rear end has now reached its final stage, using smooth sheet metal; a towing hook has also been installed. (*Ford Co. ref. 77476-23*)

the elevation of the gun barrel. For the first time, this report put forward recommendations for all opening parts of the drivers' positions to be locked, together with suggestions for keeping the two roof shutters horizontal so they would not put pressure on the hinges.

On 30 November, a second report, in collaboration with delegates from the Cavalry, Tank Destroyer Command, the Armored Force, and Army Technical Services, started defining the basic points of how the production-series vehicles were to be fitted out. The racks had to be ready

for the .45-caliber machine pistols; the turret had to have a lot of external fittings for hooking equipment on, and a series of U-shaped hooks for a tarpaulin; a terminal box for radios was missing; and the cartridge supply for the small arms had to be stowed in boxes of 50.

Chains for the six wheels also had to be supplied and foot rungs on either side of the hull replaced by racks for stowing three anti-tank mines each. The final important decisions were made at the beginning of December 1942, putting an end to "finalizing" the M8.

Above, both: The various hand tools (shovel, etc.) are now in their definitive positions, some no longer on the sides of the vehicle. Even though the images are early production photos, the anti-tank mine racks are in place (also used as steps). The rear fender skirts are now mounted on hinges and those at the front have been re-designed. (*Ford Co. ref 77476-34*)

Above: Germany, 1945. To make access even easier, the crew of this 629th Tank Destroyer Battalion M8, assigned to the 5th Armored Division, has welded a step to the front of the body. (*U.S. Army Signal Corps*)

had been rejected. On 17 December, installing a pioneer-type compass in the driving position was made compulsory, but since delivery deadlines could not be reduced to less than six months, production would have to go ahead without that accessory. On 19 December, official confirmation was received that the periscopes and most of the turret's roof were to be removed.

Before the vehicle was put into production, various other changes were made, some of them simple, others essential; the hand tools were installed differently and the siren was moved from the left to the right side of the forward hull.

The rear side fender design was modified, the skirts mounted on hinges to make maintenance and repairs easier. In order to avoid the front wheel arches getting clogged up, the size of the side skirt was reduced. The fuel tanks had to be self-sealing, like on half-tracks, in case they were perforated by small-arms fire. Carrying out some of these modifications took time but did not prevent the first examples from coming off the production lines. These examples were extensively tested in the months following the first deliveries.

Crew Drill

Taking on the new combat vehicle came with strict rules which the crews had to follow. The Light Armored Car therefore came with a new crew

On 9 December, the first of three meetings took place to define the quantities of steel needed for the program and to include them in the National Emergency Steel Program so there was enough, Ford being in charge of defining its monthly long-term needs.

On 15 December, the factory was informed its request for getting rid of the dashboard ammeter

HOW THE FIRST 159 EXAMPLES OF THE M8 LIGHT ARMORED CARS WERE DISTRIBUTED FROM THE FACTORY	
General Motors Proving Ground (GMPG), Milford, Michigan (2 vehicles delivered)	
Chicago	#15
Saint Paul	#10
Tank Arsenal Proving Ground (TAPG), Utica, Michigan (2 vehicles delivered)	
Chicago	#17
Saint Paul	#46
Aberdeen Proving Ground (APG), Aberdeen, Maryland (4 vehicles delivered)	
Saint Paul	#6, 8, 42, 44
Chester Tank Depot (CTD), Chester, Michigan (88 vehicles delivered)	
Saint Paul	#36, 38, 40, 48, 52, 54, 56, 58, 60, 62, 64, 66, 68, 70, 72, 74, 76, 78, 80, 82, 84, 86, 88, 90, 92, 94, 96, 98, 100, 102, 104, 106, 108, 110, plus 55 even numbers between #112 and #224 except #208.
Lima Tank Depot (LTD), Lima, Ohio (60 vehicles delivered)	
Chicago	#7, 9, 11, 13, 19, 21, 23, 25, 29, 31, 33, 35, 37, 39, 41, 43, 45, 47, 49, 51, 53, 55, 57, 59, 61, 63, 65, 67, 69, 71, 73, 75, 77, 79, 81, 83, 85, 87, 89, 91, 93, 95, 97, 99, 101, 103, 105, 107, 109
Saint Paul	#12, 14, 16, 18, 20, 22, 24, 26, 28, 30, 50
Ford River Rouge Plant, Dearborn, Michigan (2 vehicles delivered)	
Saint Paul	#32, 34

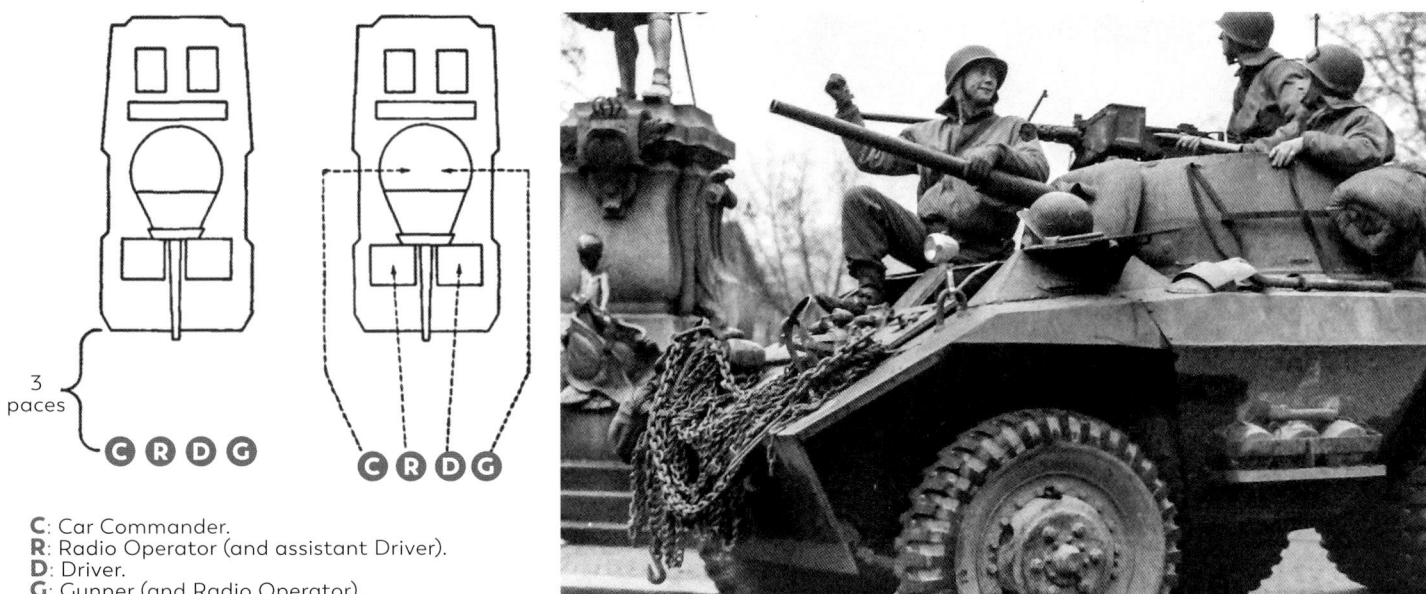

3 paces

C R D G   C R D G

C: Car Commander.
R: Radio Operator (and assistant Driver).
D: Driver.
G: Gunner (and Radio Operator).

training handbook—Field manual FM 2-6. It described all the procedures the men had to adopt. In the M8 context, only the first paragraphs are of interest.

1) Training and exercises underlined how important it was to always carry out the correct signals quickly. Strictly adhering to standard procedures during

Above left: Extract from Field Manual FM 2-6: the diagrams for the "getting into the vehicle" drill. On the left, where to stand when on parade; on the right, where the men sat inside the vehicle. It was certainly theoretical, but it did enable the men to acquire necessary automatic reactions.

Above right: In the town of Moers, Germany, 8 March 1945, with elements of the Ninth U.S. Army. Placing the wheel chains in such a manner has several advantages: they are at hand; they do not need stowing away; and, especially, they can be used as a step to get to the drivers' positions. (*U.S. Army SC-201818*)

Right: The tested vehicles were delivered without any accessories or machine guns and, for some of them, even without the 37-mm gun.

Below: When it came out of the factory in Chicago on 24 May 1943, #15 was transferred to the General Motors Proving Ground at Milford, Michigan, on 20 June, where it is seen in this photo. (*U.S. Army Signal Corps*)

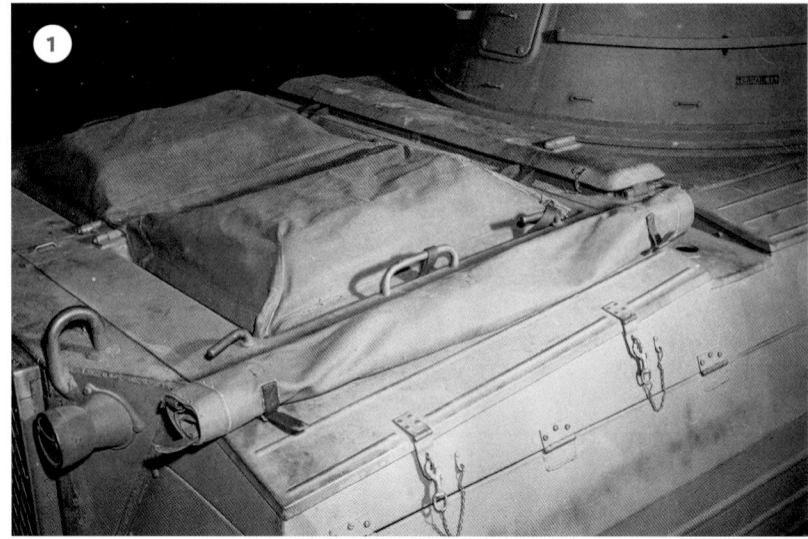

the instruction period ensured the crew worked properly as a close-knit team in combat situations.

2) The crew of the M8 consisted of four men: car commander, radio operator (and co-driver), driver, and gunner/radio operator. The car commander could be an officer or a non-commissioned officer.

3) On parade, the crew lined up in a single rank. The car commander stood three paces in front of the right-hand wheel, facing the vehicle. The radio-operator, driver, and gunner, in that order, stood on the right of their commander, in a closed rank.

4) Each of the crew members took his place as indicated on the diagram.

The manual continued with all the instructions on how to use the vehicle and all the special vocabulary used when the car commander issued orders to the other crew members. The installation diagrams also showed how useful the lateral steps were.

The front not having any of these accessories, the crews improvised to enable the driver and co-driver to get in more easily.

Tests and Validations

A big testing program was launched at the beginning of April 1943. From 8 April to 2 June, 159 production-series M8 Light Armored Cars were shared out among various centers to approve the validity of the project. The vehicles were taken in hand not only to be tested but also to be used to start training and refining tactics. The number of vehicles available was slightly more than the number received by the Army since deliveries had begun three months earlier. All early production vehicles were thus allocated to the tests.

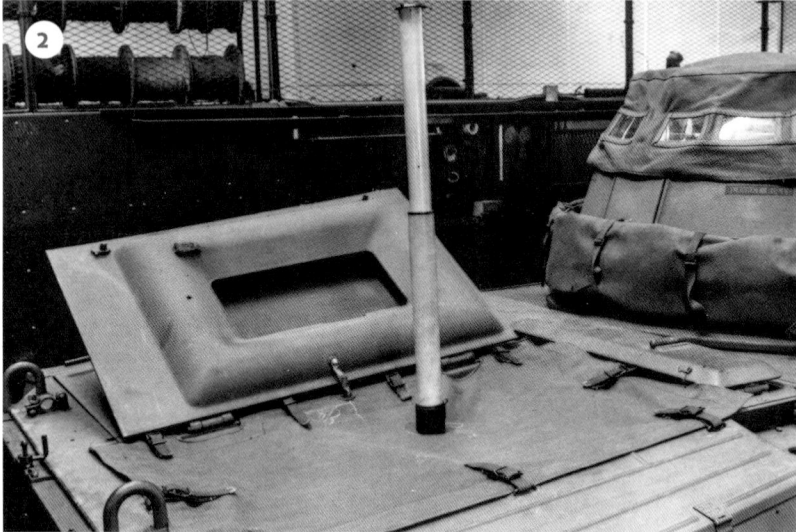

1). Installing the winter protection on the engine compartment covers. They prevented snow from getting into the engine cooling system and also protected from the cold. (*Ford Co. ref. 78274-1*)

2). When the engine was at the right temperature, and there was no longer any risk of freezing, a tarpaulin was fitted in place of the right-hand cover. It protected from bad weather and allowed a tube to be slipped through for the exhaust gases from the interior heating system to escape. The lower portion of the turret had a sausage-like roll to protect the swivel mechanism and the turret was protected by a tarpaulin canopy with little windows. (*Ford Co. ref. 78274-8*)

3). For protecting the rear in winter, a removable panel reduced the intake of air and created chicanes which blocked snowflakes. (*Ford Co. ref. 78274-2*)

HEATING

1) Installed behind the driver's seat, the heating unit vented its exhaust gases under the combat post's floor towards the engine compartment. (*Ford Co. ref. 78274*)

2) Passing under the fuel tank, the piping comes up alongside the battery to end in a metal tube with several sliding elements, each inside the other. (*Ford Co. ref. 78274-11*)

3) It was held in place by a collar attached to a bracket fixed to the wing nuts holding the battery. This extendable unit was topped by a filter to protect it from all sorts of foreign bodies. (*Ford Co. ref. 78274-11*)

4) This flexible metal pipe was covered with asbestos cloth where it risked causing damage due to its radiant heat. (*Ford Co. ref. 78274-9*)

Great lessons were learnt from these trials which lasted until September. A lot of weak points were discovered in the steering and the suspension. Though all these little problems could be dealt with on the spot, they did mean certain modifications had to be made on the production lines, like fitting more suitable shock absorbers. The factory, which was excluded from the tests, wondered why there were so many breakdowns. The Ordnance Department admitted, begrudgingly, the methods certain specialists in charge of the tests had applied had strayed from the limits set for the M8.

Unfortunately, it was soon obvious the design and, above all, manufacture of the turrets could be improved. A report dated 24 June 1943 and signed by First Lieutenant J. R. Murray, stated that more than half of the turrets revealed welding problems. Sixty-seven units delivered to the Chester Tank Depot were repaired on the spot and another 21 delivered to the Lima Tank Depot were sent to Chester for rework. Not all the vehicles were treated in the same way, some suffered a particular fate.

Vehicles #28 and #30 were redirected to the Tank Destroyer Board at Camp Hood, Temple, Texas; #24 and #26 went to the Cavalry Board at Fort Riley, Kansas; and #12, 14, 16, 18, 20 and 22 went to the Armored Force Board. All were built in the Saint Paul factory and came from Lima Tank Depot, Lima, Ohio.

Vehicle #6, intended for the Aberdeen Proving Ground, was diverted on 8 April to Ford Highland Park, Michigan; it was lent to the manufacturer for about 10 days and then replaced by #46. These two vehicles were used for finishing the technical handbooks being prepared.

Vehicles #12, 14, 16, 18, 20 and 22, which had been handed over to the Armored Force Board, suffered particularly during the Fort Knox tests. They each clocked up 4,000 miles off road. The Ordnance Department decided to recondition them and allocated a budget of $30,000 to carry out this upgrading. The Chester Tank Depot performed the work. The supplier of the special fuel tanks, Firestone Tire and Rubber Co., recovered the six used fuel tanks and sent them to its laboratories for analysis. Six new tanks were supplied free.

Vehicle #32 left the Ford Saint Paul factory on 12 March and was directed to the Ford River Rouge complex for hot and cold climate condition studies. It profited from the tests already carried out on another of the group's vehicles, Armored Car T17. The installation of two petrol heaters, one in the fighting compartment and the other in the engine compartment, the fuel being taken directly from the vehicle's fuel supply, was considered. A tarpaulin kit was also made with insulating covers for the engine hoods, and a protective tarpaulin with small windows for the turret. These were ultimately not standardized.

Vehicle #50 went to the Supply Branch at Camp Cook, California, on 20 March for the shooting of the first instructional film, called "Mechanized Patrolling," about the M8 Light Armored Car. Vehicle #11 was ordered to join it.

A lot of tests were planned for the first 159 M8s, the most marked being the 11.00×18 sand tires fitted to #7 and the 5,000-mile road endurance tests for #9 and #46 together with all the structural studies, carried out before and after the trials. A lot of thinking was done, even if none of it was really put into practice, like replacing the original JXD powerplant with the more powerful JXLD version or using self-blocking differentials. Only a single variant of the front suspension, with triangles and torsion bars, was tested, but not before 1944. These 159 test vehicles[1] were not sent to any units; they were all used as training vehicles and one of them ended up being used as a target.

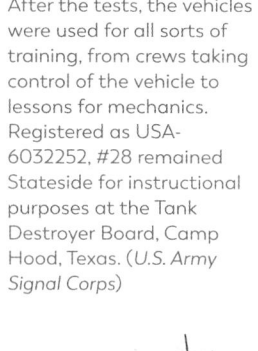

After the tests, the vehicles were used for all sorts of training, from crews taking control of the vehicle to lessons for mechanics. Registered as USA-6032252, #28 remained Stateside for instructional purposes at the Tank Destroyer Board, Camp Hood, Texas. (*U.S. Army Signal Corps*)

---

1 Where these 159 vehicles went is detailed in the Appendix on pp. 155–158.

Early Production, Serial #10 in Detail

Light Armored Car #10, taken from the first series of 159, is the perfect example of an early production vehicle. Its Ordnance Serial Number, 10, defined it as the fifth vehicle of the series produced, numbers 1 to 5 having been the two T22s, the two T22E1s and the T22E2.

Ordered as part of contract No. W-374-ORD-1744, it received the registration number USA-6032234, even though this is not visible in the accompanying photographs.

Above: Twenty-seven long months had passed since the 37-mm Gun Motor Carriage was launched and before the M8 Light Armored Car was put into production. Photographed in one of the assembly halls, #10 is on display with all its driving position panels closed. The storage racks for the anti-tank mines, also used as steps, are the early models; they have not yet received the central support. (*Ford Co. ref. 77441-13*)

Right: Opening the armored hatches gave the crew better visibility for driving, but also exposed them to the weather and the enemy. (*Ford Co. ref. 77441-14*)

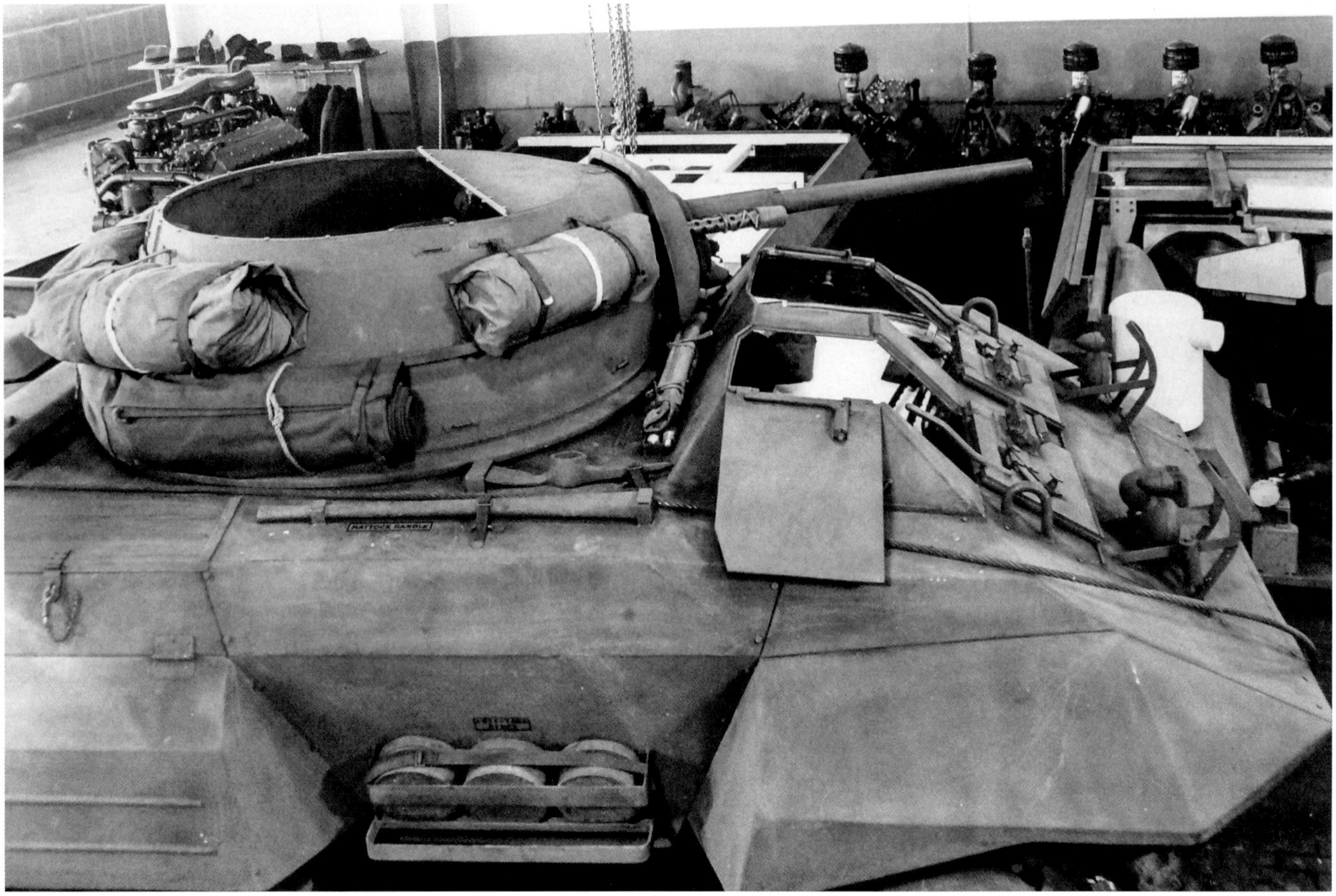

THE VEHICLE'S FEATURES

Above: The pickaxe and its wooden handle are stored on the right side of the body, near the turret. The towing cable, according to the regulations, passes between the handle and pickaxe. Around the top of the turret are four lots of two rolled-up blankets (one roll per crew member) covered in canvas. At the bottom of the turret is a big 12×12-foot tarpaulin for building a temporary shelter. (*Ford Co. ref. 77441-15*)

Right: In the right-hand lockers is a tripod for setting up the .30-caliber machine gun on the ground, the tool bag, an M2 decontamination sprayer, two machetes, a pouring spout, a jack with its crank handle, a wheel brace, a turret tarpaulin, a ventilator belt, a dynamo belt, a pair of asbestos gloves, and a grease pump. (*Ford Co. ref. 77441-18*)

In the left-hand lockers are the chains for the six wheels, another M2 decontamination sprayer, a canvas bucket, and a haversack holding all the technical handbooks: FM 23-7, FM 23-65, FM 23-30, FM 23-81, TM 9-743, and SNL G-176. (*Ford Co. ref. 77441-17*)

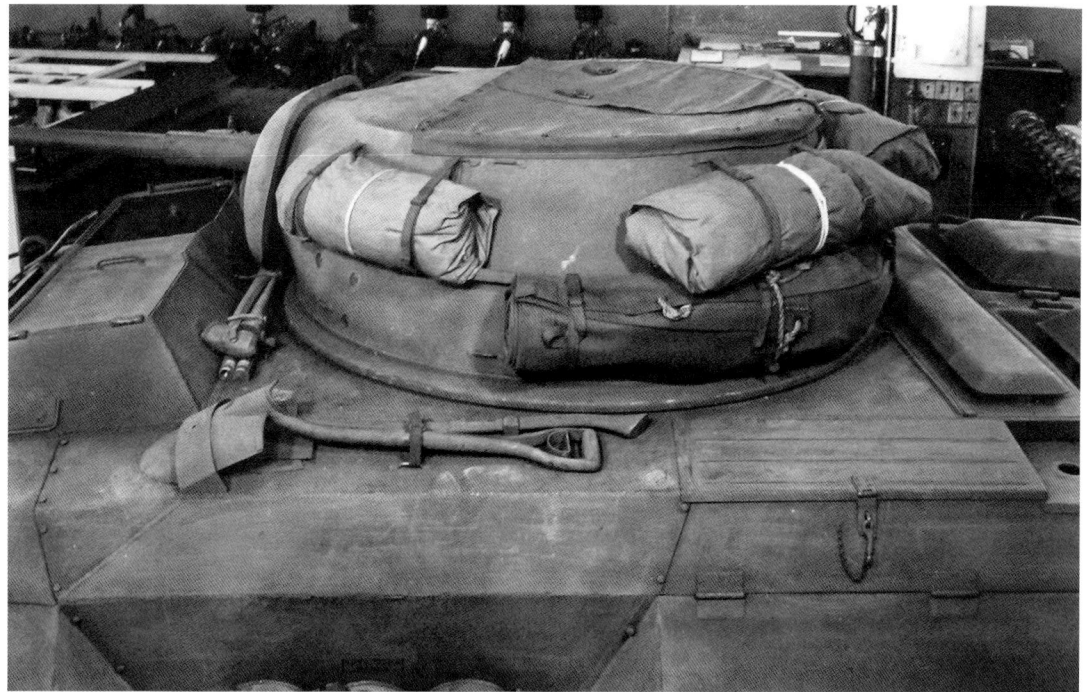

On the left side of the body, near the turret, are the shovel and the axe. Just behind the drivers' positions is the aerial cover and the swab kit for cleaning the gun barrel. The turret opening is covered by a tarpaulin held in place by a series of clips incorporated into the upper crown. (*Ford Co. ref. 77441-16*)

Taken on charge in March 1943 with 14 other vehicles, it was taken from the Saint Paul factory and delivered to the General Motors Proving Ground, Michigan, to undergo mechanical and stability tests. Before leaving the Ford factory, it was set up as the subject of a photo documentary which was distributed to the heads of the nine test centers given the job of validating the M8 Light Armored Car program. It shared something special with nine other identical vehicles from the beginning of production: none of them had any retaining brackets for keeping the driving position's roof panels horizontal.

THE DRIVING POSITION

As seen here, photographed from the turret, the driver's position on the early M8s was at its most basic. Only during production, and as and when needed, as defined by the Ordnance Department, was it filled with a variety of accessories. (*Ford Co. ref. 77441-9*)

With the seat base placed directly on the floor, the driver's and co-driver's positions were uncomfortable; space being very cramped, the driver did not have anywhere to put his left leg, so a footrest was welded to the body, but this only partly relieved the discomfort. To make a vehicle with the lowest outline possible, the body was positioned as low as possible and needed an internal bulge for the front differential to clear properly. (*Ford Co. ref. 77441-11*)

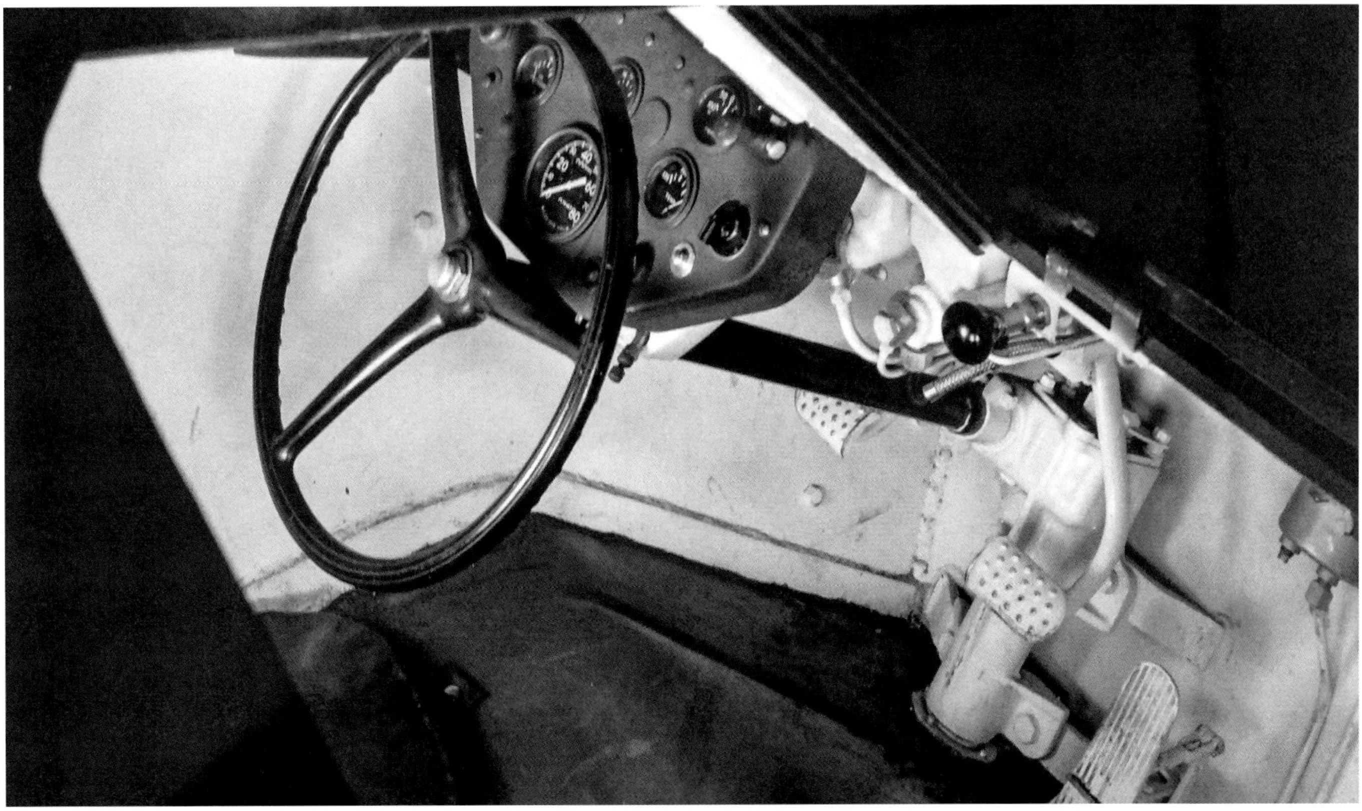

THE DRIVERS' POSITIONS

Above: All the controls and dials have been gathered on the instrument panel behind the steering wheel. (*Ford Co. ref. 77441-6*)

Right: The handbrake lever was positioned horizontally behind the dashboard; under it, the hydraulic control block receives the impulses from the accelerator and clutch pedals and transmits them to the rear part of the vehicle; on the left, the clutch master cylinder; in the center, the brake master cylinder; and on the right is the priming pump. Welded to the partition on the left of the steering column, is the left leg footrest. (*Ford Co. ref. 77441-12*)

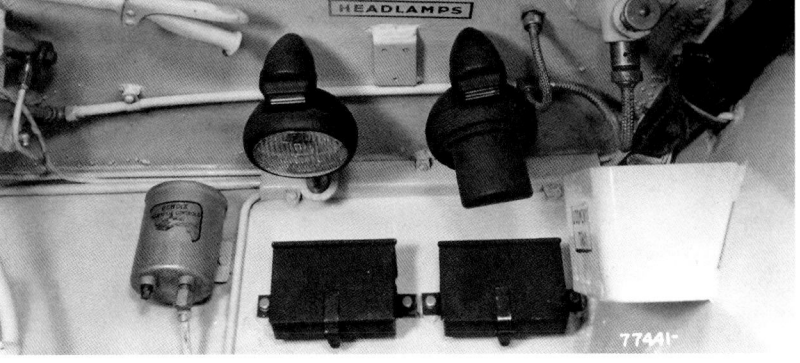

Right: The sides of the co-driver's position were fitted out to take his individual weapon (an M1 carbine), two spare vision blocks and two mountings for spare/unused headlights. On the right of the blackout light are the electrical connections for the siren and the front headlight, and the headlight socket locking mechanism. (*Ford Co. ref. 77441*)

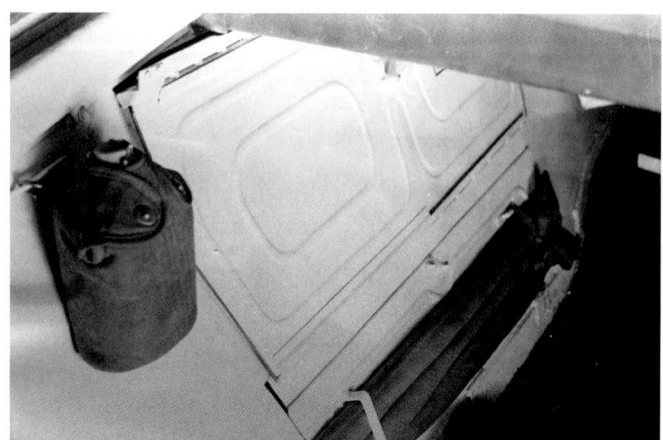

Above: Behind the driver's seat were installed the water jerry can, the toolbox and gun spare parts, and the on-board fire extinguisher. The upper nook on the left-hand side body bulge is empty, waiting to be filled by a radio set; the lower part houses a small box of .30-caliber ammunition.

Bottom left: Behind the co-driver's seat in the right-hand side bulge, the racks are protected by doors, the lower part by a shutter opening downwards. On it is a bag containing the three signal pennants. (*Ford Co. ref. 77441-8*)

Bottom right: In the absence of a radio set, the upper part houses a case of sixty-five 37-mm shells, directly within reach of the car commander whose job it was to load. The lower part was used for stowing 24 K-rations. (*Ford Co. ref. 77441-7*)

THE TURRET

Left: The right side of the turret was reserved for the car commander, in charge of operations, loading the gun and using the co-axial machine gun. On his right was the reserve of eight shells directly within reach, with another eight behind him. The left side of the turret was occupied by the gunner who was in charge of rotating the turret, aiming, and firing the gun; he was also the loader for the machine gun, whose belts passed over the 37-mm gun. (*Ford Co. ref. 77441-22*)

Left: The internal design of the turret easily accommodated all the improvements requested when the prototype T22 was tested. The seat supports were linked to each other at their base and extended under the gun to form a rigid structure. Though simply built, the turret still needed to be handled delicately. It did not have a basket rotating with it, so it was imperative for the two crewmen to stay seated, rest their feet on the structure and coordinate their movements well. No chance here of putting their feet on the floor and risk causing an accident when the turret swiveled. (*Ford Co. ref. 77441-20*)

Below: This angled view from the rear of the vehicle reveals the overall layout of the turret. (*Ford Co. ref. 77441-23*)

Above: Training the drivers was done in a group; they had to learn how to handle their new mounts and how to adapt to the new reconnaissance tactics. (*U.S. Army Signal Corps*)

The M8s Come to Life

There were no slack periods between the first series, the large-scale trials, and initial deliveries to the units. All the training centers continued the evaluations, all the schools carried out their own tests, and the first training courses started for the crews and maintainers.

Below and facing page, top: This M8, without any of its equipment, is carrying several hundred pounds of lead ballast on each side. Training can thus take place under conditions as close to combat weight as possible. (*U.S. Army Signal Corps*)

The vehicles were prepared and, in order to reach their proper combat weight, they carried ballast, because it was absolutely vital to know what the new equipment's limits were. For some, a layer of sandbags on the floor helped to simulate a theoretical load; for others, lead weights held in place by straps welded to the structure gave a good impression of how the vehicle would hold the road, how its suspension reacted, and how its steering was affected.

Improving the Turret

All these maneuvers, all this training, gathered an untold amount of data and reports for the Light Armored Car. Some information was already known after the Ordnance Department's tests; others continually improved the project. The great number of structural faults found in the turret during the preliminary trials led to its basic design being re-thought. On 8 July 1943, the Ford factory was allowed to make a turret out of welded plates 0.625 of an inch thick to replace the turret cast with sides 0.70 inches thick. The trials were not conclusive, so it was decided to modify the existing turret slightly by strengthening its sides, thickening them by 0.05 inches to make them three-quarters of an inch thick.

In a report dated 8 June 1943, Ford was ordered to modify the position of the firing sight on the gun mount.

Vibrations were transmitted through the sight; some users complained moreover about it being too close to the cannon. The Armored Force Board ordered it to be moved 2 inches to the left. This decision meant the structure of the gun mantle and the turret had to be redesigned and changed.

The plans were only available on 10 August for forwarding to the various parties.

On 4 October 1943, after delivering the gun carriage, the Miller Printing Machinery Co., told the Ordnance Department it had already made 3,700 examples of the sight located near the cannon and that it could only start to make the new design once it had exhausted its stock.

The Engineering Manufacturing Branch of the Tank Automotive Center recognized this proposition's worth and the finalization of this important change was perfectly coordinated.

Getting the front and rear suspension right remained a major worry. The first step in improving this was to change the shock absorbers. The assembly lines received a revised and less-flexible version of the shock absorbers, dated 23 July 1943. This change was made on the vehicles with even series numbers from #660

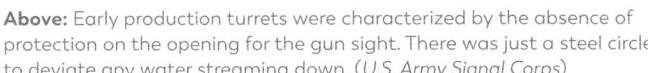

Above: Early production turrets were characterized by the absence of protection on the opening for the gun sight. There was just a steel circle to deviate any water streaming down. (*U.S. Army Signal Corps*)

Below: The first 3,700 gun carriages produced had the sight near the cannon. (*U.S. Army Signal Corps*)

Above: Very quickly, a small, curved hood was fitted to protect the end of the gun sight; when this was done is not known. (*Ford ref. 78274-7*)

Below: On the later models, there was an increased gap between the gun and the sight's protection. (*U.S. Army Signal Corps*)

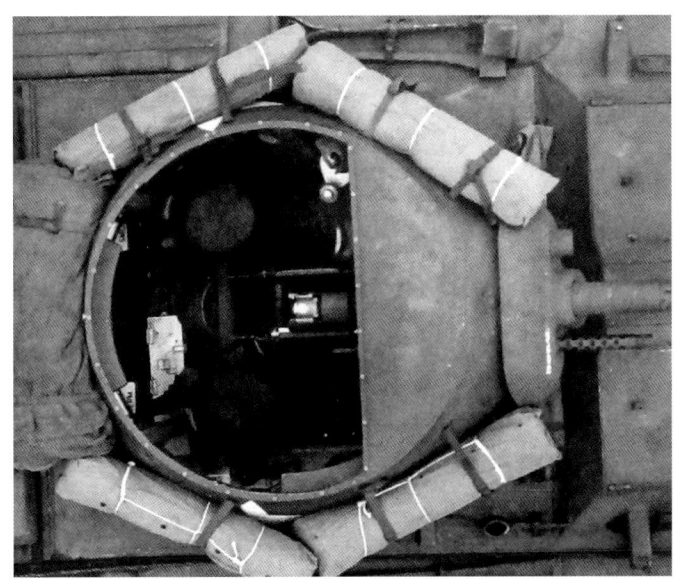

onwards, and with odd numbers from #469. The order was also given to modify all the vehicles still present at Ford and those already owned by the Army—560 examples. Although the shock absorbers were very quickly changed on all the production vehicles, the comfort they provided was only relative. The front differential especially suffered from this change which, with time, left it more fragile, though this did not seem to worry anybody. The differentiation between even and odd builder's numbers was to be found throughout M8 production; this was due to the organization of the assembly lines, the even numbers coming from the Saint Paul factory and the odd numbers from Chicago.

The M8A1

The Ordnance Department ordered another prototype to be made, based on the T21 Light Armored Car put forward in November 1942 by Studebaker, which had been considered inadmissible at the time.

The M8A1 Project, based on the M8, was therefore begun in September 1943. Ordnance Serial #393 was to be used as the test bench and modified. It came off the production line in July and all the front workings were modified so they could take independent wheels and torsion bars.

Development was long and difficult. The trials were carried out at the Aberdeen Proving Ground a year later and ended in April 1945. The first evaluation report, dated 3 April 1945, reported a distinct improvement in driving feel, describing the M8A1 as more pleasant to maneuver than the M8. This report's specific conclusions mentioned the lack of structural strength as a major reason for all the failures suffered during the trials. These structural deficiencies were caused by using forged parts instead of cast parts. A second report, dated 18 June 1945, noted that, after the defective parts were changed, the M8A1 version was certainly superior to the standard M8. But hostilities had ended in Europe more than a month earlier and the project was abandoned for good.

Top and middle left: The development of the new suspension with independent wheels on the M8A1 required the vehicle to be modified considerably. The traditional differential was replaced by a spider shaft transmission and the standard shock absorbers were replaced by Houdaille parts. (*Ordnance*)

Bottom left: A typical example of the second life of an M8 of the first series. Transferred on 29 April 1943 from the Saint Paul factory to the Chester Tank Depot for its trials, it was handed over to a unit after those trials, the 5th Cavalry Reconnaissance Troop of the 5th Infantry Division receiving it in March 1944. (*ACME ref. W717164*)

PREPARED
BY L.T.D.
11/11/43

SCR 506
SCR 508

Mid-production, Serial #2940 in Detail

Light Armored Car #2940 was the perfect example of a vehicle from the middle of the production series. Its Ordnance Serial Number 2940 defined it as being the 2,945th production-series vehicle built.

Ordered as part of Contract No. WD-374-ORD-1744, it was given the Registration Number USA-6035764 but this is not visible on the photographs. Taken on charge in November 1943, it was sent to the Lima Tank Depot, Ohio, where it was prepared on the 11th before being sent to the Ordnance Operation–Engineering Standards Vehicle Laboratory in Detroit for testing and photographing from all angles on 25 February 1944.

Above: In this side view, two stenciled notices can be seen on the body: the first confirms the vehicle was prepared by the Lima Tank Depot ("Prepared by L.T.D. 11/11/1944") and the second defines the two on-board radios, an SCR-506 and an SCR-508. (*Ordnance ref. 550-551*)

Left and following page, top: This front three-quarter profile gives a good idea of the vehicle's fluid lines. It was not by chance that the British nicknamed the vehicle "Greyhound." (*Ordnance ref. 553-554*)

Below left: The front remained identical except for the positioning of protruding rests on the sides of the casemate to keep the drivers' panels horizontal and locked in place when open. (*Ordnance ref. 546*)

Below right: This is a first-generation rear; the fender panels are smooth and the exhaust pipe ends under the fender on the right hand side. This configuration is consistent with M8s with serial numbers between 1 and 3907, except for those with an odd number between 1361 and 3907, which had the second-generation rear. (*Ordnance ref. 548*)

Opposite: With time, variations in the internal painting were encountered and are impossible to identify. Was it due to decisions on the assembly line or some other directive? On #2940, the turret accessories (including the seat frames) were painted Olive Drab; on others, like #10 described earlier, they were white. (*Ordnance ref. 552*)

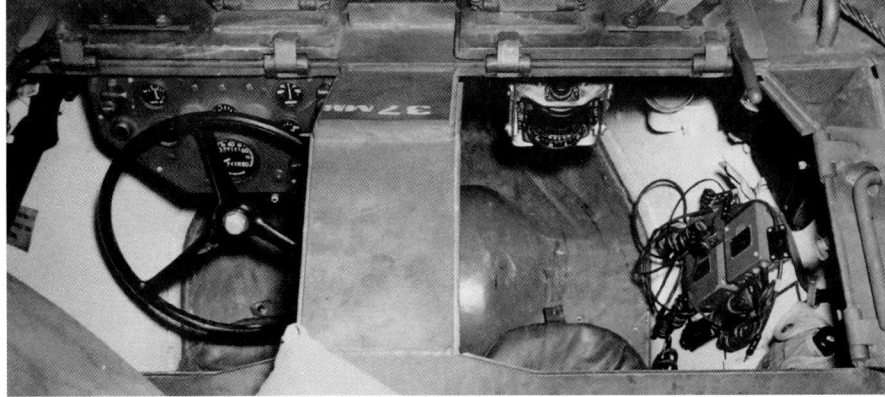

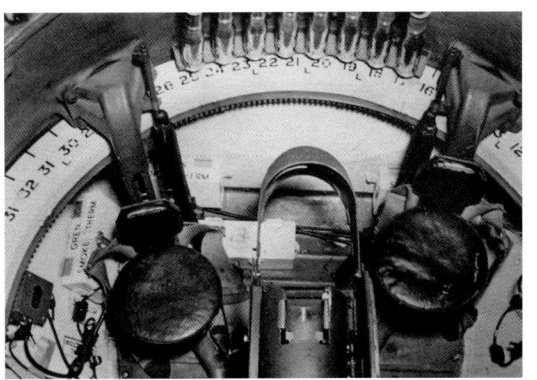

Above: This view of the drivers' position shows the position of the intercom on the right, and the compass which equipped the vehicles from #1000 (for the evens built in Saint Paul) and from #843 onwards (for the odds built in Chicago). The instrument panel is from the second generation and remained in use until #7836 and #8113 respectively. (*Ordnance ref. 560*)

Above: Apart from the differences in the shades of paint for the accessories, the inside of the fighting compartment continued to fill up with various items, like the intercom boxes. The vehicle's four seats were now fitted with a strap as a safety belt. (*Ordnance ref. 557*)

M8 Armament

Apart from the crew's individual weapons—carbines and grenades—the main armament of the M8 consisted of a 37-mm gun and a .30-caliber machine gun, both mounted on an M23 Gun Mount, inherited from the M3A1 and M5 Light Tanks.

As the testing carried out from April to September 1943 had identified several flaws in the turrets of the 159 vehicles put through their paces, modifications were needed which led to the M23A1 Gun Mount.

GUN AMMUNITION

There were six different configurations of 37-mm ammunition for the M6 gun (*Extracts from TM 9-1904*)

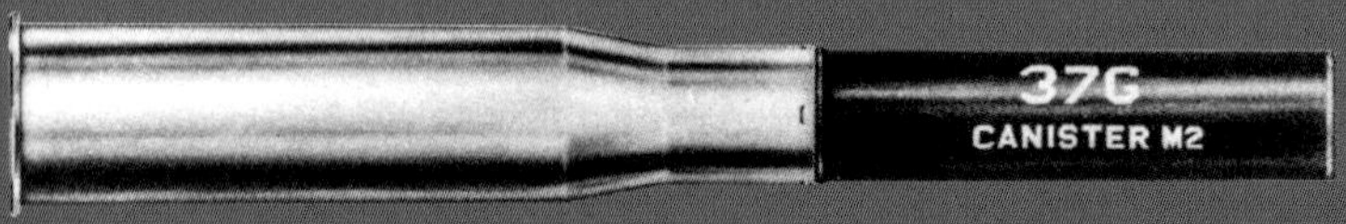

Canister, Fixed, M2, 37 mm (*Grapeshot, anti-personnel*)

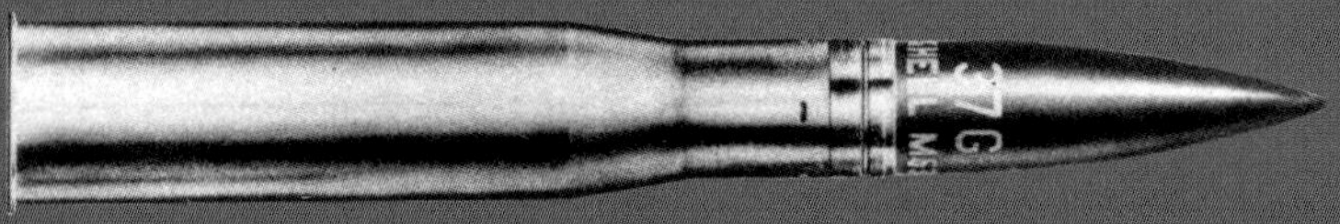

Shell, Fixed, HE, M63, with Fuse, B.D., M58, 37 mm (*High explosive*)

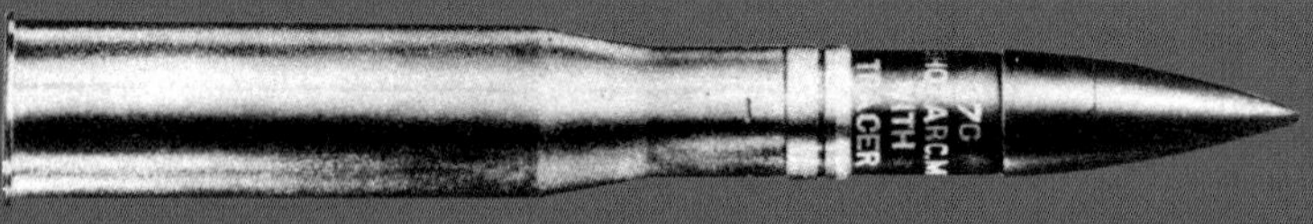

Shot, Fixed, A.P.C., M51, with Tracer, 37 mm (*Armor-piercing*)

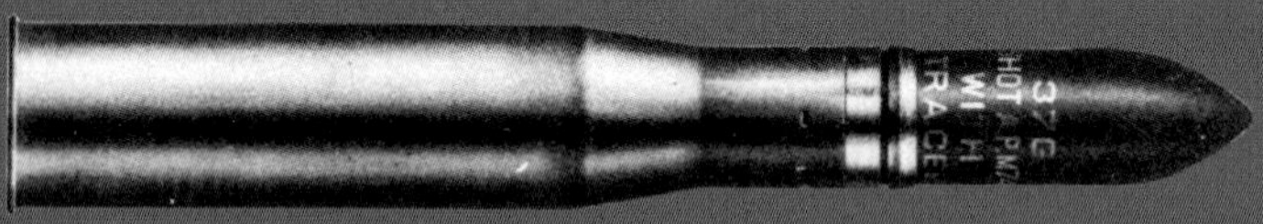

Shot, Fixed, A.P.C., M74, with Tracer, 37 mm (*Semi armor-piercing*)

Shot, Fixed, T.P., M51, with Tracer, 37 mm (*Armor-piercing training round*)

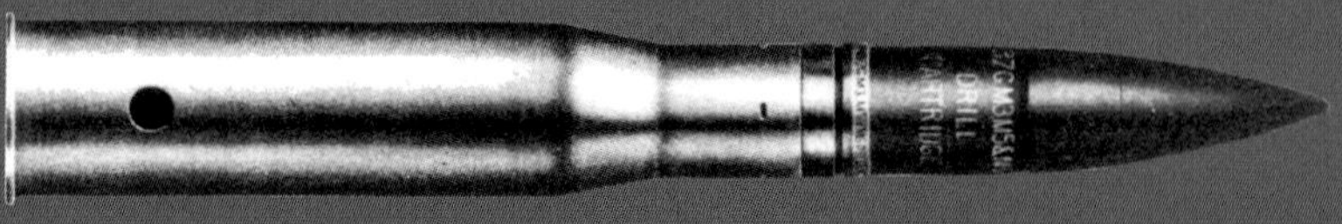

Cartridge, Drill, M13, 37 mm (*Inert training round*)

The M6 37-mm Anti-Tank Gun

This gun was the consummate version of the M3 gun (mounted on a wheeled carriage) adopted in 1938. Fitted into the 360° swiveling turret, its clearance was -10° to +20°. Its total length was 82.5 inches, and it weighed 185 lbs, accessories included. Given the turret's compact dimensions, it was impossible to replace the cannon without using a trapdoor at the back of the turret. Six shell models were specifically designed for the M3, M5 and M6 variants of the 37-mm gun.

- Grapeshot shell (Canister, Fixed, M2, 37 mm). Its head contained 1,220 steel balls. It weighed 3.49 lbs, with a range of between 150 and 200 yards, and velocity of 2,500 ft/sec. Its head was painted black with inscriptions in white.

Below: An example of how shells were stowed loose in the lower right-hand rack, when a radio set was fitted in the upper part. Safety inside the vehicle was seriously compromised as a result. (*Ordnance ref. 558*)

- Explosive shell (Shell, Fixed, HE, M63, with Fuse, B.D., M58, 37 mm). Its head was painted olive drab with yellow inscriptions. It weighed 3.13 lbs, had a maximum range of 9,500 yards at 2,600 ft/sec.
- Armor-piercing shell (Shot, Fixed, A.P.C. (Armored Piercing Cap), M51, with Tracer, 37 mm). Its head was painted black with white inscriptions. It weighed 3.48 lbs and had a maximum range of 12,850 yards at 2,900ft/sec.
- Armor-piercing low-penetration shell (Shot, Fixed, A.P., M74, with Tracer, 37 mm). Used against weakly armored vehicles like half-tracks. It was useless against tank armor. Its head was painted black with white inscriptions. It weighed 3.34 lbs and had a maximum range of 8,725 yards at 2,900 ft/sec.
- Armor-piercing exercise shell (Shot, Fixed, T.P. (Target Practice), M51, with Tracer, 37 mm). It was painted blue with white inscriptions.
- Handling shell (Cartridge, Drill, M13, 37 mm). Inert, its head was painted black with white inscriptions.

Even though the M51 shell's ballistic performance was impressive, its armor-piercing ability has to be moderated: at 500 yards, it could penetrate 2.1 inches of normal armor inclined at 30°, or 1.8 inches of special armor; at 2,000 yards it penetrated 1.4 inches and 1.3 inches respectively.

The standard issue for the M8s was 80 rounds, split into three lots: two lots of eight shells inside the turret and 64 in a rack situated in the upper part of the right-hand body bulge. The shells in the rack were removed when the vehicle had to carry two different radios. To avoid running out of ammunition, some crews simply filled up the rations rack with shells, all piled on top of each other, without any precautions to avoid an accidental detonation.

It was only in mid-1944 that two new solutions were found: the first consisted of a new rack (36 shells) under the righthand radio; the second was to modify the original rack to become a vertical block for 43 shells.

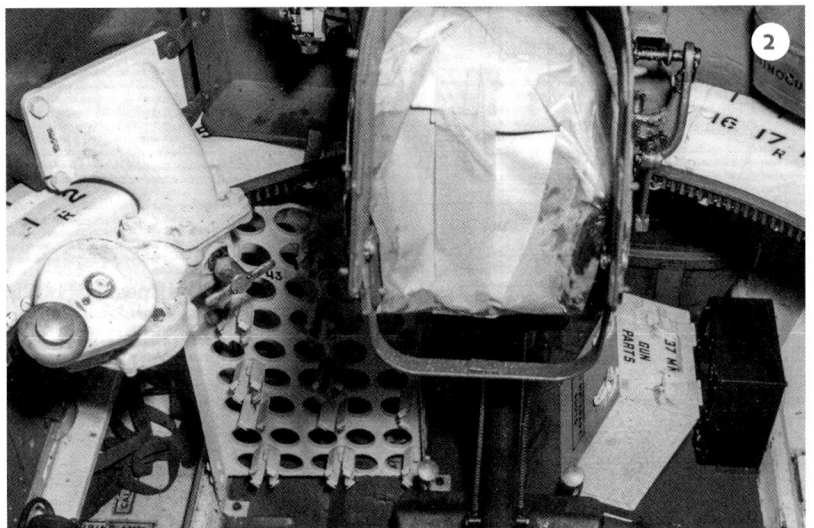

1. An intermediate solution with a new 36-shell rack at the bottom of the right-hand body bulge. (*Ordnance ref. 1514*)

2. The original rack modified, and re-positioned, to hold 43 shells. (*Ordnance ref. 1875*)

3. An alternative solution with a box containing 20 shells squeezed between the turret's two seats. (*Ordnance ref. 1876*)

This was placed in the position kept for the gun toolbox, the fire extinguisher and the water jerry can. An additional 20-shell container was also used, squeezed between the turret seats.

The M1919A4 .30-caliber Machine Gun

The classic .30-caliber machine gun was mounted co-axially to the right of the gun and could be removed for use on the ground on a standard tripod, itself stowed away in the side locker, above the rear right-hand fender.

As the weapon was so close to the cannon it could not be fed normally. The tray for the ammunition case was fitted to the left of the cannon and the belt was fed over the main weapon through a curved chute. Fifteen hundred rounds in cloth C3951 belts, stocked in D44070 cases, were stowed in the lower part of the left-hand body bulge.

The HB M2 .50-caliber Machine Gun

An AA weapon was only later supplied to the M8s. Adding such a weapon to so small a vehicle was not easy. Space had to be found for stowing its accessories and a mounting had to be created from scratch. It had 1,000 rounds, split among 10 cases stowed wherever there was space, depending on the crew's preferences; its heavy tripod was strapped to the right-hand rear fender.

Above, both: Photographed at the Aberdeen Proving Ground on 16 November 1943, this M8, registered as USA-6032268, bore the Ordnance Serial #44. It was the test vehicle for the M49C Circular Ring Mount. (*U.S. Army Signal Corps*)

Right, both: The D67511 Folding Pintle Bracket, designed by Ford for the AA .50-caliber machine gun, in position at the rear of the turret. In the second photograph, its ground tripod can be seen strapped on top of the rear right-hand locker. (*Ford Co.*)

There were three ways of installing it: a ring mount designed by the Ordnance Department, a Ford-designed mount, or the do-it-yourself mounts crews made up in the field.

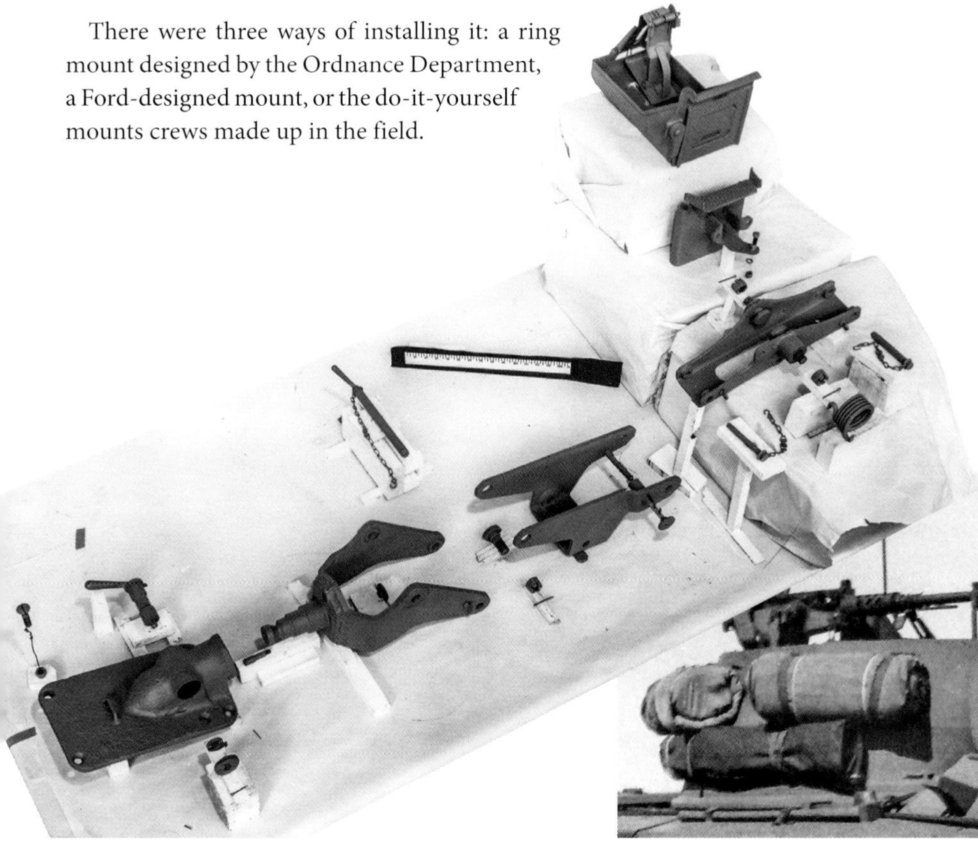

Left: An exploded view of the elements making up the D67511 Folding Pintle Bracket. (*Ford Co. ref. 81425-17*)

Below: When travelling, the machine gun was lowered by unlocking the cradle to mid-height. The barrel rested in a bracket welded on top of the turret. (*Ordnance ref. 1503*)

Bottom: The Belgian–Dutch border on 8 September 1944. The first vehicle from C Troop, 113th Cavalry Reconnaissance Squadron, First U.S. Army, patrols on the lookout for snipers. This shot is a typical example of American propaganda because the .50-caliber machine gun has not been fed with any rounds. (*U.S. Army SC-194099*)

The M49C Ring Mount

This practical solution only appeared in the second half of 1943. Taking up the widely used system on trucks, of a ring fitted with a small mounting on wheels moving through 360°, the Ordnance Department created the M49C Ring Mount, which was fitted to the turret with three supports. This solution was not unanimously approved:

- The men had to climb over it to get to their posts in the turret.
- It was practically impossible to use the machine gun when the vehicle was in combat mode and when the crew had to serve the cannon.
- The gunner's position was very unstable because he had to perch on the two seats.

Despite it being standardized late, at the end of 1943, 2,645 examples of this mount were ordered from the Trakson Company, under contract No. W-21-ORD-1547. Administrative reception reported 2,411 were delivered in 1943, followed by 501 more in 1944.

The Folding Pintle Bracket D67511

Ford created its own version of the .50-caliber machine-gun mount. It kept the space free above the turret. The rear trapdoor panel, which enabled the gun to be changed, was replaced by a panel that could take the weapon's forked pivot. The only basic difference between the two panels was the number of fittings, four for the basic version and six for the improved version.

Tested by the Cavalry Board, the design of the Folding Pintle Bracket was fully approved. It was lighter than the Ring Mount elements; it was easier to swivel through 360° because this was done from a single, central point; it did not take up any space on the top of the turret; the machine gun could be

Above left and left: U.S. troops have just crossed the Franco–Belgian border on 3 September 1944. M8 Light Armored Cars are passing through Rongy, 7.5 miles south of Tournai in Hainault Province. There are several similarities on the vehicles of this unidentified unit: the front fenders have been removed and a tarpaulin has been placed on the lower hull to hide the unit markings. (*U.S. Army SC-194013-194380*)

Above: At Kinsweiler in the Rhineland, 21 November 1944, this M8 with a homemade ring mount is the 31st machine from A Troop, 17th Cavalry Reconnaissance Squadron, Ninth U.S. Army. The 17th and the 15th made up the 15th Cavalry Group, Mechanized. The 17th operated in this region in support of the 30th Infantry Division. (*U.S. Army SC-196666*)

Above, right: The 82nd Armored Reconnaissance Battalion, 2nd Armored Division, enters Saint-Sever-Calvados on 3 August 1944. Both M8s have homemade ring mounts. (*U.S. Army SC-332086*)

used at the same time as the cannon and a man other than the crew members could fire it from outside the turret.

Homemade Ring Mounts

As the M49C Ring Mount was relatively rare, like the Folding Pintle Brackets, crews in the field used their initiative and inventiveness. Recovering M36 Ring Mounts from trucks with tarpaulin-

covered cabs became an obsession. The GIs cut the supporting legs, adapted them and welded them to the turrets, each ring being unique.

Right: Adapting heavy ring mounts from the trucks was not the only solution used in operations. Very often, simple mounts were welded on the rear, the sides, and even on the top of the turret. On this M8, crewed by 1st Lt Donald F. Engle and Cpl Henry Nilken in the turret, and Sgt Leroy F. Ammons and Pvt John J. Wisniewski in the drivers' positions, a .30-caliber machine gun has been fitted on the right using elements from an M1917 heavy machine-gun mount. (*ACME ref. A-27*)

Above: This three-quarter view shows the final look the Ordnance Department wanted to give the vehicle. As they arrived rather late, the upgrade kits were almost never used in the field during the conflict. (*Ordnance ref. 1497/1498*)

Late Production, Serial #8325 in Detail

Serial Number 8325 was the perfect example of a vehicle from the end of the M8's production run. It was the guinea pig for the new external side lockers, the protective wind shields for the drivers' posts, and the storage locker on the front of the hull; it then went to the Ordnance Operation–Engineering Standards Vehicle Laboratory in Detroit to be tested and photographed from all angles on 10 August 1944.

Right: A bird's eye view with all details quite visible, including the ground mount for the .50-caliber machine gun and its stowage cradle on the top of the turret. (*Ordnance*)

Above and below: Left and right profiles: #8325 was fitted with prototype side lockers (where the star is painted). These were to change again with a step being installed to assist access to the top of the vehicle. (*Ordnance ref. 1495/1494*)

Above left: The new turret mantle with the sight moved away from the cannon. (*Ordnance ref. 1493*)

Above right: The rear with the new exhaust pipe outlet through the right-hand-side fender plate, the location of which was identical to the later M20 Armored Utility Cars. (*Ordnance ref. 1496*)

Above: The updated gun mount elements with the sight moved away from the gun and the second-generation turret swivel mechanism. (*Ordnance ref. 1511*)

Right: A view of the turret base with the canvas chute for collecting the spent cases from the gun, the footrest for the gun crew, and the two foot-activated triggers. (*Ordnance ref. 1513*)

The T69 Multiple Gun Motor Carriage

In the long line of studies carried out on the T22 leading to the adoption of the M8 Light Armored Car, and just after the M20 Armored Utility Car (or, in military parlance, Car, Armored, Utility, M20) was created, the Ordnance Department authorized a study to be made for a Multiple Gun Motor Carriage using the same chassis and mechanicals.

Above: On 29 April 1943, the Ordnance Department drew up a specification for the T69 Multiple Gun Carriage and proceeded to officially take it on charge at the Aberdeen Proving Ground. (*Ordnance ref. 82383*)

The project was strongly supported by Tank Destroyer Command which wanted a ground-support and AA-defense vehicle. Its common base with the M8 and M20 meant streamlined maintenance, repairs and logistics together with a proven ability to keep up with the forces it was supporting. The engineer's primary studies, carried out with Ford, showed that, using the features of the M45 Quadruple Gun Mount as a base, it would be possible to have a group of four .50-caliber machine guns swiveling through 360° with clearances between -10° and +85°. Trials with a ballasted M45 showed the mechanicals and vehicle structure could bear the weight.

Right: With its armament at maximum elevation (+85°), it is clear to see the space between the two lower weapons has been taken up by ammunition boxes. (*Ordnance ref. 82582*)

The W. L. Maxson Corporation started designing and building the new turret prototype in November 1942. As far as Ford was concerned, it modified an M8 chassis following Maxson's instructions, adapted the vehicle's fittings and changed it from 32 to 24 volts for 50 amps. This enabled the vehicle, when the engine was running or it was moving around, to power the turret, which was all electric. The factory fitted a small independent generator for when the engine was stopped.

For no particular reason, Army Ground Forces decided the trials would be carried out by Anti-Aircraft Command and excluded Tank Destroyer Command from the entire validation process of the T69 Multiple Gun Carriage.

At the end of April 1943, after having wedded the turret and the vehicle, it was delivered to the Aberdeen Proving Ground. From 12 to 16 May 1943, the T69 underwent its first running and firing tests. Although the results were clearly

Above: Placing templates on the ground, each attached to the fenders via a stylus, enabled the identification of stresses and vibrations as they transmitted to the rest of the vehicle when the guns were fired. (*Ordnance ref. 83269*)

Middle left: All the turret controls were grouped together in a single control panel underneath the sight. (*Ordnance ref. 83272*)

Bottom left: Reloading was carried out manually from the turret. (*Ordnance ref. 83647*)

Above: August 1943, the modified T69 has returned to the Aberdeen Proving Ground for a new series of tests. (*Ordnance ref. 90824*)

Left: The main visible changes concerned the rear of the turret with larger cutouts for maximum weapon elevation and a new, higher-profile front (*Ordnance ref. 90825*)

satisfactory, there were some criticisms: inadequate vision for the gunner; empty cartridges piling up at the bottom of the structure; lack of comfort of the seat; and a risk of preventing the turret from swiveling. The vehicle was driven back to Maxson for improvements and modifications.

The T69 returned to the Aberdeen Proving Ground in August 1943, having been modified as requested, including a new profile for the front of its armor. It was finally driven to Camp Davis, North Carolina, for a study comparing it with the M16 AA Half-Track, itself also equipped with four .50-caliber machine guns. At the end of March, a very short report sealed the fate of the T69.

- The T69 was superior to the M16 Half-Track because of the crew's protection. The crew was reduced (three men instead of five). It was also more suitable for engaging ground targets.
- The capacity of the T69 to carry enough ammunition was considered inadequate. It was, above all, considered to be too cramped.

In conclusion, in March 1944, Anti-Aircraft Command recommended the evaluations be stopped and the project scrapped.

Left: September 1943, the vehicle has suffered during its mid-year tests. The turret armor has been removed as have the incorporated ammunition boxes. (*Ordnance ref. 91625*)

3

Car, Armored, Utility, M20

The T26 Prototype

In accordance with the memo dated 25 March 1942 regarding the evaluation of the T22 Gun Motor Carriage, the predecessor of the M8, further development was pursued using production-series vehicles.

As Tank Destroyer Command desired, attention now focused on creating a command vehicle and a troop transport and/or supply vehicle. The movement order was to start with an M8 from which the turret, as well as all the structure between the drivers' positions and the engine compartment, had been removed to refit the interior for the two required functions.

The command vehicle was designated, in military parlance, Car, Armored, Command, T26 and was to carry two passengers who could work on maps and signals easily. The transport vehicle was designated Carrier, Personnel-Cargo, T20 and was built to carry six fully equipped men, or about 3,000 lbs of supplies.

This view, greatly touched up by Ford technical services, represents the final version of the T26 Armored Command Car. Note that the base is an M8 with the front touched up by hand and the superstructure grafted on. (*Ford Co.*)

A single vehicle was made available for the study and, if it satisfied all demands and the factory went with it, it would be called "Car, Armored, Utility, T26," which is what happened.

It is interesting to discover what the initial impressions of the vehicle were. In a report dated 15 March 1943, First Lieutenant J. R. Murray recorded how the new vehicle handled.

"... I left by train on Wednesday 10 March at 10 o'clock and arrived at Aberdeen on Thursday 11 March at about 10 o'clock. The T26 was already there, having arrived there from the Ford factory by road by its own means on 10 March."

"On Thursday 11 March, in the morning, the vehicle was checked and, in the afternoon, it took part in the first ballistic trials with a .50-caliber machine gun on the M32 Ring Mount delivered with the vehicle. Two hundred rounds were fired from different positions around the ring, and with various elevations, in order to evaluate bullet dispersion. The firing showed the horizontal dispersion was low but that the vertical dispersion was too much to be considered satisfactory.

"On the morning of Friday 12 March, the vehicle was prepared to be demonstrated to Army Ground Forces (AGF) representatives. Colonel Wales from HQ/AGF, Captain Parquette from the Development Branch, and Captain Perraine from the Office of the Chief Ordnance came directly from Washington to inspect the vehicle. They were mandated by all sections of the Army to approve or reject the new vehicle. Another firing trial with the .50-caliber machine gun was made before Colonel Wales himself drove the test track at Aberdeen. To his great satisfaction, he was unable to fail the vehicle on that very muddy course and was very satisfied with its general performance.

"He later agreed, in the name of all sections of the Army, that the vehicle could be adopted. Captain Parraine prepared the contracts and promised to speed up the adoption procedure by referring to Colonel Wales's verbal approval. Captain Parquette supported this procedure and proposed the designation Car, Armored, Utility for the new vehicle.

"I left Aberdeen at 4:30 pm on 13 March and returned to Detroit at 10 o'clock on Sunday 14 March."

First Lieutenant J. R. Murray

The tested and approved vehicle was given the Ordnance Serial #1. After this session at Aberdeen, it went for three 2-week sessions of trials, first at Camp Hood (Texas) under the responsibility of the Tank Destroyer Board, then at Fort Riley (Kansas) to be examined by the Cavalry Board, and, finally, to the Armored Force Board at Fort Knox. For these evaluations, the task was made much easier because the structural and mechanical bases had already been studied at length when the M8 Light Armored Car was developed.

Above: Apart from the modification to the combat compartment, the new project was identical to the M8 design which the Army began to take on charge in March 1943. (*Ford Co. ref. 77533*)

Below: At this stage of the evolution, no accessories have been fitted. Among the other similarities to the M8, we can see the mine racks and the step have not been linked by a middle reinforcement bar. (*Ford Co.*)

The First M20 Armored Utility Cars

The program progressed very quickly as the Ordnance Department wanted the M20 to catch up with the M8, but the order of progress was not entirely logical.

While the T26 prototype had barely finished its testing, Ford could easily commence production. As early as April 1943, the T26 was standardized under the designation "Car, Armored, Utility, M10." But this designation was not suitable because M10 already referred to another armored vehicle, the 3-inch Gun Motor Carriage of Tank Destroyer Command, so the designation for the armored car was changed to M20.

Above: The Armored Force Board at Fort Knox, Kentucky, testing a first-generation Armored Utility Car. (*AFB ref. 6728*)

Left: Presentation of a fully equipped vehicle, with its left side lockers open. The contents are the same as the M8 Light Armored Cars. (*AFB ref. 6723*)

The first four examples were ordered on 31 May 1943 to be transferred as early as possible for the initial stages of final approval. At the beginning of July, Ordnance Serial #2 headed off to Rouge Plant at Dearborn, Michigan, for a comparative analysis of its design at the Ford Motor Company's Engineering Department.

Ordnance Serial #3 was driven to Fort Riley, Kansas, to undergo tests and evaluations by the Cavalry Board.

Ordnance Serial #4 was exhibited at Ford Highland Park, Michigan, and was used for publishing technical handbooks. Ordnance Serial #5 left for the Aberdeen Proving Ground for driving tests and, above all, to solve the problem of excessive bullet scatter when the machine gun was fired, as already noted. Dispatched between 6 and 8 July, these four vehicles were the very first to come off the production line and to be taken on charge.

Meanwhile, the first reports on the T26's six weeks of tests and inspections were sent in by Tank Destroyer Command, the Cavalry Board, and the Armored Force. Colonel K. L. Cummings, in the name of the Chief of Ordnance, answered the various remarks in a report dated 16 May, but not without having first sought the advice of the manufacturer on certain points. Several solutions had already been found for the M8.

1. The problems of the steering column vibrating had been corrected by fitting a rubber block and springs.

2. A footrest for the driver's left leg had been fitted during production. This new support consisted of a plate welded on the side of the body directly to the left and slightly to the rear of the clutch pedal.

3. The system for blocking the vision slots did not seem to have any major defects. It was identical to what was used on the M8, of which a large number had already been built and no defects found.

4. The question of an adjustable footrest for the co-driver had already been answered. It would have meant a complicated mechanism. Adding a safety handle was envisaged but no footrest of this type was ever installed.

5. The design of one or several handles remained on hold. This study was delayed because the navigation compass had to be fitted; its mounting had to be installed beneath the armor

Top: The layout on the right flank was also the same. Without the turret, on which certain accessories could be attached, extra attachment points were fitted to the engine compartment cover panels. (*AFB ref. 6722*)

Middle: A three-quarter view of the same vehicle. (*AFB ref. 6724*)

Above: All the crews trained intensively to learn how to use their new mount. (*U.S. Army Signal Corps*)

Right: A friendly encounter between an M8 from the Third U.S. Army and an M20 from a French unit, at Autun in the Saône-et-Loire Department. (*U.S. Army Signal Corps*)

Below: A column moving through the Belgian town of Saint-Hubert on 9 September 1944. Strangely, the stars have been painted on the inside of the drivers' roof panels. (*U.S. Army Signal Corps SC-195959*)

plate of the forward hull, directly in front of the co-driver.

6. At this stage of the development, there were no handles in the main compartment. It is difficult to understand why these were requested since no specific needs had been mentioned.

7. Safety belts were planned for the two seats in the drivers' positions.

8. No changes were made to improve the space above the drivers' heads. The request for a 2-inch bulge in the panels was not upheld. The driver's position was the same as the M8 and no unfavorable report had been received on this point. Moreover, modifying the panel profiles meant costly modifications.

9. It was not possible to enlarge the rack door panels on the body for cases of .50-caliber ammunition because of how the benches and the radio shelves were positioned.

10. The on-board equipment list stated that a 5-gallon jerry can of water was placed at the rear of the left-hand side panel with the fire extinguisher placed on the other side of the crew's compartment. The extinguisher bracket was such that it could take a second jerry can of water if required.

11. A solution to make the steering column or the steering wheel movable, thus making access to the driver's position easier, was under study. Until then, no satisfactory arrangement had been found. The study continued and, as soon

Right: On the village square of Beaumont in Belgium, Gen J. Lawton Collins, commanding VII Corps, has stopped his armored utility car so he can get his bearings on the map. (*U.S. Army Signal Corps SC-377508*)

Below: In the Baden-Baden region of Germany, the 11th Armored Division is waiting its turn to cross the Rhine. In the middle of the traffic jam are vehicles of all types—including two captured German Sd.Kfz 251 Half-Tracks covered with American markings. Two M20s are seen patiently waiting. (*U.S. Army Signal Corps SC-202215*)

as a suitable design was found and tested, it would go into production.

12. The lever to engage the front-wheel drive was strengthened and the whole production series was suitably equipped.

13. The left-hand racks had already been redesigned to ensure the SCR-506, 608 and 618 radio sets were properly installed. It was certainly better to keep the basic design of the left-hand rack with which the model vehicle had been equipped, i.e., the mounting of the SCR-610 supports with a hinged metal access panel to close it off. This access panel had, however, been lengthened and made removable and, in order to install the SCR-506 or the larger 608, the panel was removed, as were the SCR-610 supports, and a flat radio shelf installed in this rack.

14. A slightly bigger pulley had been installed on the generator to avoid over-speeding and to increase belt life. An auxiliary generator, a copy of the generator in standard vehicles, was under study and proposed as a kit. It would supply an extra 50-amp recharge rate or a total of 100 amps for the vehicle.

15. At the same time, the Signal Corps was studying the problem of interference when two radio sets were used at the same time.

Studying the report was interesting because the answers formulated for several issues were almost more important than the specialist's questions in charge of the tests. They gave an insight into the improvements carried out and the intellectual steps which led to solving the problems brought up. One can also deduce, given how pointless some of the questions asked were, that the teams which had analyzed the T22 a year earlier were not the same as those who were now studying it,. One can also assume, even though it was not said implicitly, that the base used for building the T26 was none other than the second T22 prototype, cast aside during the M8 study. The remarks about the footrest on the left of the clutch pedal, and the transfer lever for the front drive axle, confirm the T26 was not built on the base of an M8 but on that of the second T22.

Two great problems remained to be solved: finding a solution for increasing the electrical power, so both radios could be used at the same time; and finding the right ring mount for the .50-caliber machine gun.

Early Production, Serial #181 in Detail

Utility Armored Car #181 was the perfect example of a vehicle from the beginning of production. Its Ordnance Serial defined it as the 180th production-series vehicle, #1 having been attributed to the prototype T26.

Ordered under the one and only contract, W-374-ORD-1744, it was given the registration number USA-60110953. Taken on charge in August 1943 with 204 other similar vehicles, it was initially delivered from the Chicago factory to the Ordnance Operation–Engineering Standards Vehicle Laboratory in Detroit, Michigan, for various tests.

Before leaving its test site, it was the subject of a photographic report, handed out to the people at the five test centers entrusted with validating the Car, Armored, Utility, M20 program. With all the other early production vehicles, it shared something very special: equipment issued in 1943 and laid out according to 1943 standards.

Above and right: Three-quarter front-right and rear-left views of M20 Armored Utility Car #181. (*Ordnance ref. 679/681*)

Right: The position of the six rolls of blankets was defined in the regulations as two rolls inside the main compartment and four outside, two strapped to the rear and one on each side. Attachments points were fitted for each. In 1943, the allocation of anti-tank mines was six per vehicle, three on each side; in 1944, it was reduced to three. In 1943, the protective tarpaulin for the crew compartment did not yet exist. (*Ordnance ref. 683*)

Middle: A bird's eye view of #181 and the layout of various accessories and equipment. The ring mount has been pushed right back to enable the four men in the combat compartment to get in as easily as possible. One can also see how the bedding rolls are positioned, as well as the tripod for the machine gun (just forward of the engine compartment). (*Ordnance*)

Below, both: #181 shows all the features common to the early production vehicles, such as the low exhaust outlet pipe (just visible where the right fender meets the hull) and the design of the rear. (*Ordnance 678/680*)

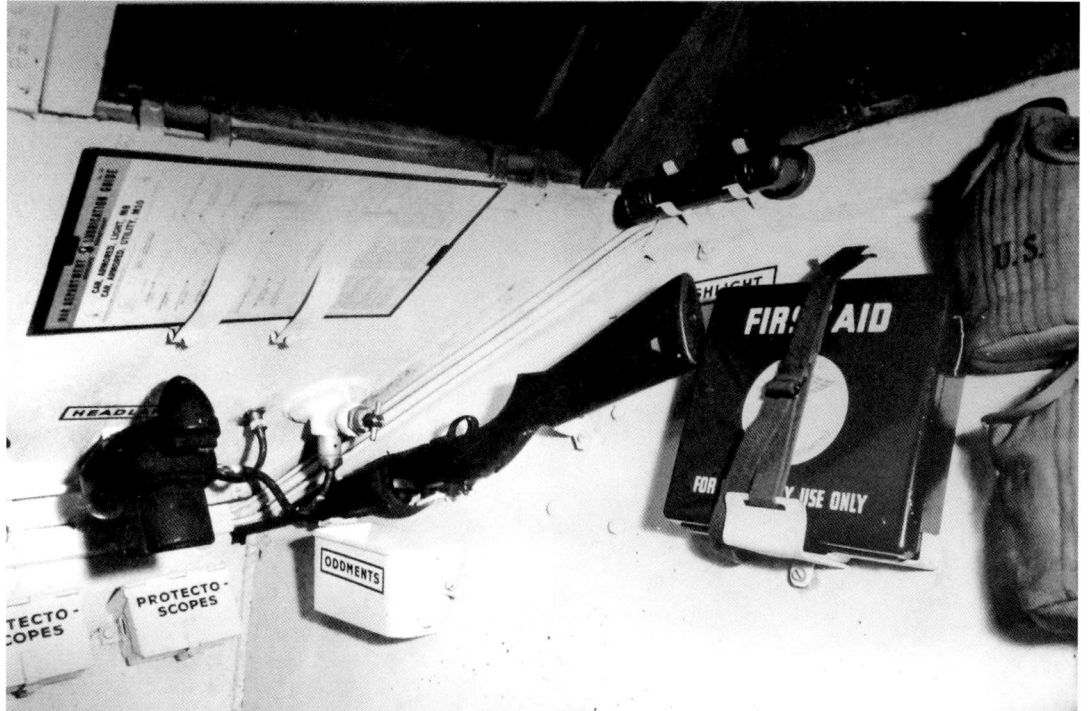

Above: The layout of the driver's position was similar to the M8. Only the positioning of certain accessories varies; the dashboard is a first-generation example with the engine compartment fire detection lights. (*Ordnance ref. 685*)

Left: The co-driver has the same elements in front of him as an M8 co-driver. The first-aid box is now on his right as are two canteens of water. Within reach are the anglehead flashlight and his carbine. (*Ordnance ref. 687*)

Development of the M20

Above: Car, Armored, Utility, M20 in an operational configuration according to the February 1944 technical manual. Prepared at the Lima Tank Depot on 28 January 1945, this example has its .50-caliber machine gun installed on the M49 ring mount. (*Ordnance ref. 2836*)

The M20 must not be compared to the M8. Although both were designed using the same base, they were employed in totally different roles.

The M20 was considered a command and logistics support vehicle, while the M8 was a combat vehicle. The .50-caliber machine gun on the M8's turret was an anti-aircraft weapon and, when necessary, could be used on the ground, but this armored vehicle's firepower consisted of its gun and the co-axial .30 machine gun. Whether its .50-caliber was accurate or not was irrelevant.

In the case of the M20, the .50-caliber machine gun, designed especially for both static and mobile ground targets, was the only heavy weapon it had, so it was important for it to be as accurate as possible. During the T26 prototype and first production-series vehicle tests, the M49 Ring Mount was noted to prevent the gun from operating satisfactorily. Many tests were carried out modifying every element; a new ring, new carriage, new fittings, and other details were tested without any real improvement.

The only progress made with firing precision came from a system called the Allis-Chalmers Type Ring which made the structure and ring

Left: The Allis-Chalmers ring mount, simply welded to the superstructure, made getting into the vehicle very difficult. Although its characteristics solved some of the firing accuracy problems, it was abandoned. (*Ordnance*)

more solid, eliminating some of the inherent vibrations. However, this structure was heavy and cumbersome, producing only marginally better results.

On 20 August 1943, Lt-Col Joseph M. Colby, in charge of development with the Ordnance Department, approved the decision to fit the M49 Ring Mount to the M20 in the factory, despite its defects.

The ammunition issued was 1,000 cartridges in 10 boxes of 100 rounds.

Five .30-caliber M1 Carbines were supplied for the crew: two in the drivers' positions and three on a rack at the rear right-hand side of the combat compartment; 500 rounds were stowed in a box behind the co-driver's seat. Regulation-wise, there was an anti-tank M1 rocket launcher (bazooka), as well as 10 M6 rockets. The weapon and eight rounds were stowed along the front side of the combat compartment with the remaining two on the rear. Three anti-tank mines were stowed on the outside, on the right or left-hand side; 10 hand grenades were placed in two boxes on the rear side of the combat compartment and six M1 or M2 smoke

Above: Another shot of the Allis-Chalmers ring mount for the .50-caliber machine gun. (*Ordnance*)

Below: A bird's eye view of the combat compartment, taken from the rear deck looking towards the drivers' positions. Even though only four men made up the crew, there was not much space. The map table, at the top in the center, is unfolded. (*Ordnance ref. 689*)

bombs were stowed under the rear seat and in the lower left-hand side of the caisson. This quantity of arms and ammunition could vary, depending on the crews' needs and various technical handbooks, the description above being the regulations as they were in February 1944.

The Combat Compartment

The central part of the vehicle was fitted out for four men and the vehicle's entire equipment, no matter what use the M20 was put to—transport, logistics support or command. Each outside bulge on the body was compartmentalized for stowing equipment or radio sets. A narrow wooden backless foldable bench was fitted along the sides.

Against the rear, a small foldable seat, for the car commander, was also used as a booster when firing the .50-caliber machine gun. At the front of the compartment, a foldable table was used as a work table and for map reading.

In the central compartment, together with the small arms and ammunition, there were four M-1910 mess kit cans (and two at the drivers'

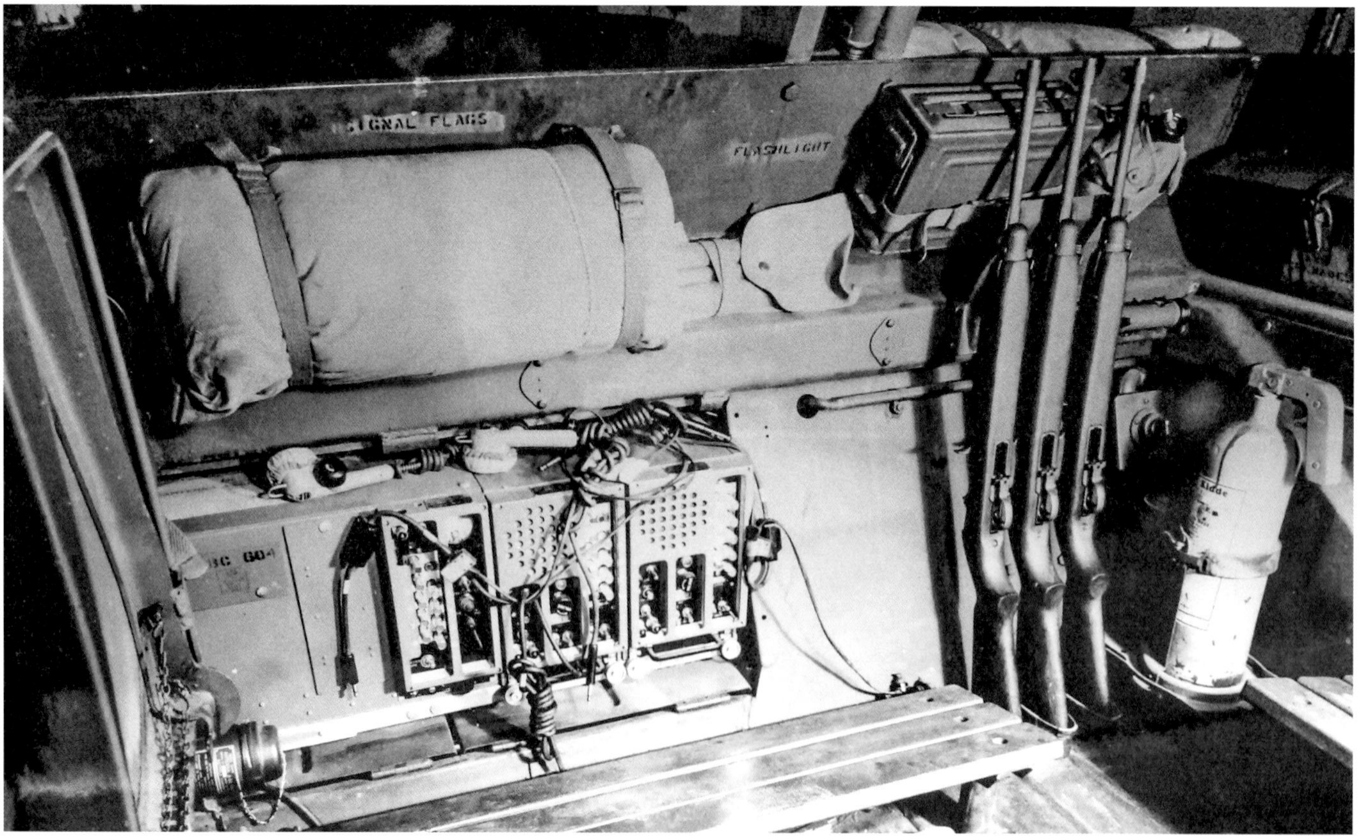

Above: Looking towards the rear, from the top of the drivers' positions. The layout of accessories, like the bed rolls on the external sides and the rear, was according to the 1943 regulations. (*Ordnance ref. 688*)

Below: Right-hand inside flank with its rack for three carbines and an SCR-506 radio set housed in the right-hand body bulge. On the left of the can and the machine-gun ammunition box are three signal flags in their covers and an individual bedding roll. (*Ordnance ref. 690*)

positions), three elbow lamps (as well as another two with the driver) and 38 combat rations stowed in the lower part of the right-hand side bulge.

On the outside were stowed the larger accessories: four of the six bedding rolls were attached two by two on the engine covers (the other two were inside the combat compartment); the protective tarpaulin for the central compartment was placed like a horseshoe to the rear; and the tripod for the heavy machine gun was stored across the front part of the rear deck.

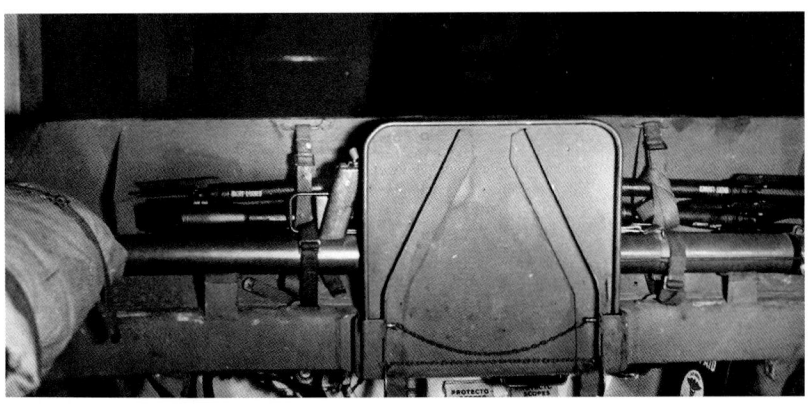

Above: A view of the small folding table mounted on the cross-member between the combat compartment and the drivers' positions. Behind it are strapped the bazooka and eight of its rockets. (*Ordnance ref. 2806*)

Left: The lefthand side with its closed radio/equipment compartments and the jerry can of drinking water. (*Ordnance ref. 691*)

The rear bulkhead angled upwards to enable the third brace of the Ring Mount to be fitted and not encroach too much upon the combat compartment. (*Ordnance ref. 2809*)

83

Late Production, Serial #3038 in Detail

Armored Car #3038 was an example from the end of the production series, its Ordnance Serial Number defining it as the 3,037th production series vehicle built, #1 having been attributed to the T26 prototype.

Ordered under contract W-374-ORD-1744, this M20 was taken on charge in January 1945, with 96 other similar vehicles from the Chicago (the only M20 plant) factory, by the Ordnance Operation–Engineering Standards Vehicle Laboratory to undergo several evaluations. Before leaving this test center, it was the subject of a photographic documentary dated 16 February 1945. Like all the other late-production vehicles, its equipment was 1944 standard.

Above: In operational configuration with all driving compartment panels open. Among the novelties, note the two hooks in place of the shackles and the horn replacing the siren. (*Ordnance ref. 2838*)

Left: The new transparent windshields for the driver and co-driver. (*Ordnance ref. 2843*)

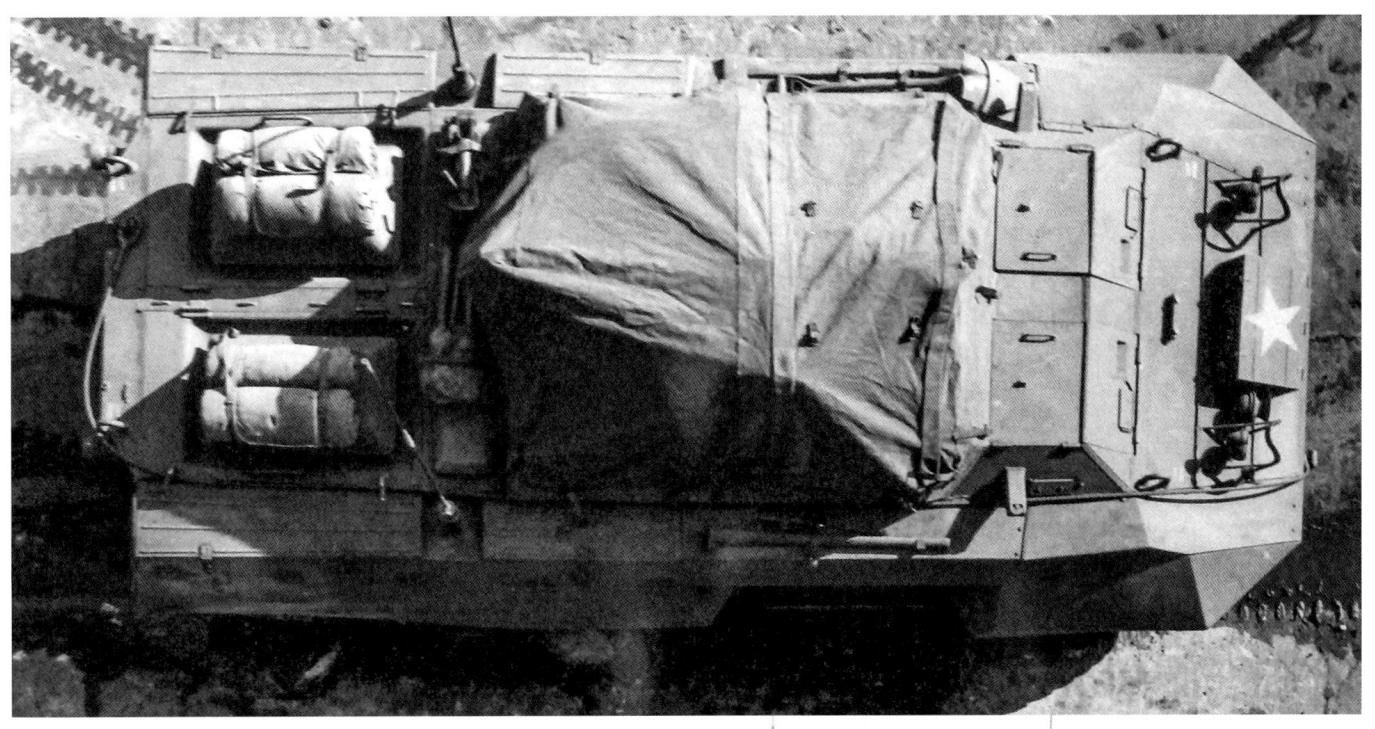

Above and right: Placing the protective tarpaulin over the combat compartment and the machine-gun mount gave the vehicle a ghostly appearance. Zip fasteners enabled the tarpaulin to be opened on the side so it was still possible to get into the vehicle from both sides. (*Ordnance*)

Left: The shackles were also replaced on the rear and the higher exhaust pipe position meant the bodywork was redesigned. The fixtures on the engine covers were used to fasten four bedding rolls of blankets. (*Ordnance ref. 2839*)

Above: Adding the large combat compartment tarpaulin changed the position of the accessories. The four bedding rolls, which had been placed on the sides and rear of the main compartment, were now moved onto the engine compartment covers. The new tarpaulin took their place and was stowed away like a horseshoe. (*Ordnance ref. 2837*)

Above: Installing the forward storage box on the front, and replacing the siren and the shackles, meant the M20 had reached the end of its development even though the new side lockers, which were to replace the mine racks, were not yet fitted. (*Ordnance ref. 2836*)

Below: This side view gives a good idea of how the layout was changed. (*Ordnance ref. 2835*)

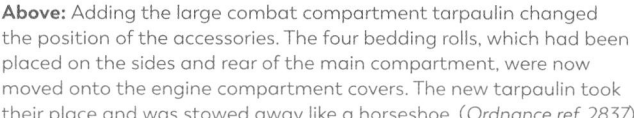

Similarities

M8 and M20 Production

Two Ford Group factories were responsible for producing the vehicles: Saint Paul, which only assembled the M8, and Chicago, which principally built the M20 and some of the M8s.

Making two similar vehicles with such different uses was a great source of conflict between the manufacturer and the Army, each keenly wanting to impose its own system of numbering and recording. The result was confusion.

The Ford System

Henry Ford's strong and authoritarian, almost rebellious, spirit was deeply anchored in his company and, all throughout production, the Mfr.

No. (Manufacturer's Serial Number) remained predominant, right up into the technical handbooks and other SNL-G[2] nomenclatures. This elementary numbering was made up of a constant, logical, mathematical series for each type of vehicle and started by including the prototypes and pre-production series vehicles. For the M20s, this was

---

2 The Standard Nomenclature List was an inventory system employed from as early as the 1930s until 1958 to catalogue all the materiel distributed by the Ordnance Department/Corps.

Under the watchful gaze of factory employees, vehicles come off the production line and immediately head to railroad platforms for loading. (*Ford Co. ref. 79590-2*)

On the way to an Ordnance depot, this train, loaded with 84 M8 Light Armored Cars, leaves the Saint Paul factory on 21 March 1944. All the vehicles have been covered with the integral protective transport tarpaulins. (*Private Collection*)

very simple because they were only built in one place: Chicago. For the M8s, this set of builder's numbers was divided between the even numbers production line (Saint Paul) and the odd numbers production line (Chicago).

TABLE OF ORDERS/DELIVERIES FOR M8 LIGHT ARMORED CARS

Number	Orders	Registrations	Serial Number	Production Order & Commentaries
2	W-374-ORD-1235	6032216 to 6032217	#1 and #2 (T3026)	Prototype, one example finished (6×6)
2	–	6032218 to 6032219	#3 and #4 (T3138)	Pre-production series vehicle T22E1 (4×4)
1	W-374-ORD-1295	–	#5	Pre-production series vehicle T22E2 (6×6)
2,188	W-374-ORD-1744	6032230 to 6034417	#6 to #2193	(T3809) All taken on charge in 1943
2,285		6034418 to 6038986[1]	#2194 to #6762[1]	(T3809) 584 received in 1943 and 1,701 in 1944
145		6038987 to 6039131	#6763 to #6907	(T3809) All taken on charge in 1944
1,382		6039811 to 60401192	#6908 to #8289	(T3809) All taken on charge in 1944
1,078		6041193 to 6042270	#8290 to #9367	(T4157) All taken on charge in 1944
1,805		60125394 to 60127198	#9368 to #11172	(T4157) 604 received in 1944 and 841 in 1945[2]
989		60132343 to 60133331	#13261 to #14249	(T18403) Orders and Canceled orders
812		60136694 to 60137505	#14250 to #15061	(T21861) Orders and Canceled orders

– 10,689 M8 Light Armored Cars ordered for only 8,528 examples received (including the five study vehicles)
– 4,299 received in 1943, 3,383 in 1944 and 841 in 1945 (plus the five study vehicles)

[1] In this series of Serial Numbers of the Ordnance and Registration Numbers, only 2,285 numbers of each were used.
[2] This series was reduced by 360 vehicles, not received.

TABLE OF ORDERS/DELIVERIES FOR M20 ARMORED UTILITY CARS

Number	Orders	Registrations	Serial Number	Production Order & Commentaries
1	—	—	#1	One example of the pre-production series vehicle
3,364	W-374-ORD-1744	60110774 to 60114137	#2 to #3365	(T7227) 1,624 received in 1943, 1,337 in 1944 and 403 in 1945
403		60131940 to 60132342	#6624 to #7026	(T18402) All taken on charge in 1945
648		60133569 to 60134216	#7027 to #7674	(T18402) 24 received in 1945 and 624 cancellations
340		60134433 to 60134772	#7675 to #8014	(T18402) Order canceled
260		60135634 to 60135893	#8015 to #8274	(T18402) Order canceled
800		60135894 to 60136693	#8275 to #9074	(T21860) Orders and Canceled orders

– 5,816 M20 Armored Utility Cars ordered for only 3,792 examples received (including one prototype)
– 1,624 received in 1943, 1,337 in 1944 and 830 in 1945 (plus one prototype)

TABLE OF ANNUAL RECEIPTS ACCORDING TO THE SUMMARY REPORT OF TANK-AUTOMOTIVE MATERIEL

Model	Number	1943	1944	1945
(T3809) M8 Light Armored Car	6,000	4,299	1,701	0
(T4157) M8 Light Armored Car	2,523	0	1,682	841
(T7227) M20 Armored Utility Car	3,364	1,624	1,337	403
(T18402) M20 Armored Utility Car	427	0	0	427
Total	12,314	5,923	4,720	1,671

TABLE OF MONTHLY RECEIPTS OF M8s AND M20s

1943												
	January	February	March	April	May	June	July	August	Sept.	Oct.	. Nov.	Dec
4,299 M8 rec'd in 1943		—	15	31	110	169	512	314	803	545	1,000	800
1,624 M20 rec'd in 1943		—	—	—	—	—	126	205	275	293	400	325

1944												
	January	February	March	April	May	June	July	August	Sept.	Oct.	Nov.	Dec.
3,383 M8 rec'd in 1944	562	468	241	223	241	234	256	243	232	234	234	215
1,337 M20 rec'd in 1944	214	193	53	48	53	32	29	83	158	160	159	155

1945												
	January	February	March	April	May	June	July	August	Sept.	Oct.	Nov.	Dec.
841 M8 rec'd in 1945	232	144	162	150	153	—	—	—	—	—	—	—
830 M20 rec'd in 1945	97	153	163	150	156	111	—	—	—	—	—	—

According to First Lieutenant J. R. Murray's notes, we know the real production in the Ford factories was:
– For the M8 Light Armored Car: 4,421 units in 1943, 3,488 in 1944 and 614 in 1945;
– For the M20 Armored Utility Car: 1,664 units in 1943, 1,412 in 1944 and 715 in 1945.

Admittedly, these figures are not of great historical significance, but they do show that manufacturing was tight compared with deliveries and that stocks in the factories was kept at a minimum.

But this list of numbers for the M8s was incomplete; there were entire blocks missing. This particularity meant there were more builder's numbers than vehicles, the SNL-G indicating numbers close to eleven thousand whereas only 8,528 body shells were built. For example, in the SNL-G, one finds that vehicles which should have borne the serial numbers between 9490 and 10186 never received a carburetor, and so, there are no identifiable vehicles for this series of 694 numbers.

These builder's numbers were stamped on the identification plate at the driver's position, in the middle of the cross-member on the rear of the vehicle, and in the middle of the upper part of the forward hull.

The Ordnance System

The Army imposed its own numbering without considering the possibility of there being a builder's number for the chassis; two lists of serial numbers were imposed, one for the M8s and another for the M20. But it was illogical for the same vehicle to be

Above: From the beginning of production, and throughout the conflict, trials were carried out with the factory. The aim was to enable the greatest number of officers to give their opinion directly to the Ford engineers, without going through official administrative channels, thus saving time. (*Ford Co. ref. 79590-13*)

Middle: A discreet off-road reconnaissance vehicle with good speed, the M8 Armored Car found its way through wooded terrain easily. (*Ford Co. ref. 79590-9*)

Left: As they had not yet been taken on charge, these demonstration examples carry no markings, including the registration number which has been masked over. (*Ford Co. ref. 79590-1*)

Above: A restoration of M20 factory number 3136-C by Mr Benoit van Pottelsberghe. On the front, this Ordnance Serial Number is found the bottom of the hull, just at the junction with the bodywork of the wings and the fenders. (*Vincent Dumont de Chassart*)

Below: Depot O-640 of the Ordnance Department near Tidworth, Wiltshire, west of London, 23 March 1944. A line-up of M8s waiting to be delivered to their units. They are a little more than two months away from the Normandy landings; the vehicles have been given their identification markings (white stars) and the figure 9 for their Bridge Classification. (*U.S. Army Signal Corps*)

defined by two numbering systems, the builder's and the Army's. A compromise was found between the two parties, adding a prefix and suffix to the military production numbers. The M8s were given the prefix "GAK" in front of their serial numbers, while the M20s, were allocated "GBK." In order to differentiate the vehicles made in Chicago from those made at Saint Paul, the suffix "C" was placed after the number.

These Ordnance serial numbers were stamped on the identification plate in the driver's position and on the four corners of the armored body.

Although this solution satisfied the military administratively, it embarrassed the Ford factories. They were forced to put these numbers on the vehicles, but they carefully avoided putting them in the various technical handbooks. For example, the catalogues of parts, SNL G-136 (M8) and 176 (M20), only mentioned the builder's numbers in defining how the variations applicable to a series of vehicles were to be used. To complicate matters, how production lines were organized at Ford's two sites meant odd and even series of numbers had

Assembly hall in the Chicago factory. These M20 Armored Utility Cars are practically complete. Only their external accessories are missing. (*Ford Co.*)

Ford personnel and the 5,000th M8. This vehicle came off the production line in February 1944. (*Ford Co.*)

Courtesy Brian McMahon

to be used. This particularity produced no logical sequence in certain cases where modifications were involved. It was a real headache which could only be avoided by going through the list of spare parts carefully so as not to get the assemblies and sub-assemblies, which were not compatible among themselves, mixed up. Although the military serial numbers were put on the armor, with the Chicago factory "C" suffix when needed, the "GAK" and "GBK" prefixes, on the other hand, were generally missing.

Production

After the long development period of a project started in 1941, with production only beginning in March 1943, everything would lead one to believe the M8s and M20s were modern vehicles from the end of the war; this was not the case when one reads the reports on the production and delivery of the vehicles.

At the end of 1943, 4,421 M8 Light Armored Cars had already come off the production lines, or more than 52 percent of the total production; as for

After the war, many countries wanted to buy the large number of surviving M8s. To satisfy all these requests, the USA drew on its reserves and sold vehicles which had been placed in mothballs. This is a partial view of a delivery of 100 M8s, ready to be loaded on 29 May 1956, in the port of Galveston, Texas, for delivery to France. (*United Press*)

the M20 Armored Utility Cars, there were 1,664 units built by the same date, which was more than 44 percent of the final number. At the end of 1944, there were 3,488 more new M8s and 1,412 M20s, leaving just 614 M8s and 715 M20s to be built in the final months of production to May 1945.

Without First Lieutenant J. R. Murray's professionalism and personal notes, taking down production and delivery figures every month, it would be impossible to establish this relationship between the data of production, delivery, and reception.

The two vehicles' entire production was punctuated by anecdotes, little stories or surprising orders. Although the Army officially ordered 10,684 M8 Light Armored Cars, it stopped its orders at 8,523 units received by the end of May 1945. Likewise, for the M20s, the Army ordered 5,815 examples but had only taken 3,791 on charge by the end of June 1945.

On 24 October 1944, the 6×6 T22 prototype and the two 4×4 T22E1s were ordered to be transferred to Fort Custer, there to be disarmed and scrapped.

Lend-Lease to the Allies for these two vehicles amounted to only 1,414 vehicles split between 1,209 M8s (689 for France, 496 for England, 20 for Brazil and four for Canada) and 205 M20s (all for France). The main American users were Tank Destroyer Command, the Mechanized Cavalry and the Armored Force; they did not want them too widely distributed to other countries.

M8 and M20 Mechanicals

Putting aside the design of the combat compartment, the two vehicles used the same engines and running gear, using identical components.

Taking the thinking further, what was pleasing with these vehicles was that certain parts were interchangeable with those already available from other manufacturers' vehicles. The Hercules JXD engine block was the same as that in the M3A1 White Scout Car, as well as the Studebaker US6 and US 6×4 trucks, or the Diamond T-614s, Federal 2Gs, or Autocar U-2044s. The universal joints, regulators, generators, fuel filter, and pump were also common, as were parts of the hydraulics systems.

THE HERCULES JXD ENGINE

1. Carburetor	19. Filler tube
2. Engine water outlet	20. Engine oil pressure socket
3. Thermostat housing	21. Clutch bell
4. Generator	22. Clutch
5. Fan belt	23. Inspection chamber
6. Ventilator fan	24. Oil level tube
7. Pulley	25. Oil level gauge
8. Fan belt	26. Engine water inlet
9. Engine block	27. Water pump
10. Starter	28. Cabling
11. Fuel inlet	29. Ignition head
12. Fuel pump	30. Pulley
13. Valve	31. Generator electric box
14. Spring	
15. Oil filter	*(Extracts from SNL-G 136/176)*
16. Coil	
17. Oil filler	
18. Engine temperature plug	

Below: View of the rear of the engine compartment of an early production vehicle where the cube-shaped thermostat housing and the Auto-Lite brand of regulator are easily recognizable. (*Ford Co. ref. 77476-21*)

Shape and Dimensions

The M8 Light Armored Car was built on an entirely welded freestanding body, made up of plates of armor and in which only the engine compartment had no bottom. The lower part of the forward hull was a half-inch thick, the top part was 0.75-inch thick, and the sides, as well as the rear, were 0.375-inch, while the top and underneath were a quarter inch. The sides of the turret were 0.75-inch thick, and the cannon mantle was 1-inch thick. The vehicle therefore weighed 14,500 lbs empty and 17,400 lbs in combat condition.

Weight distribution was 30 percent on the front axle and 35 percent on each of the rear axles.

The vehicle was 196.75 inches long, 100 inches wide and 78 inches high, with a wheelbase of 104 inches, the gap between the front differential and the second axle being 80 inches, and the rear differential 128 inches.

As for the M20, it was 68.34 inches high, just under 10 inches less than the M8; it weighed 12,800 lbs empty and had a combat weight of 15,650 lbs. It was relatively comfortable thanks to 11 semi-elliptical spring leaves and six shock absorbers, made by Gabriel, which improved road holding. For optimal driving comfort, it was recommended to inflate the front tires to 60 psi and the rear ones to 50 psi.

The Mechanicals

Using Firestone 9.00×20, 12-ply Mud & Snow Combat Tires, the Ford M8s and M20s had three axles. The front axle was an F-M8-H Model

Above: The right and left sides of the engine compartment of M8 #10. *(Ford Co. ref. 77441- 2 and -10)*

Below: The Hercules JXD engine from the end of production with all the latest modifications, like the connection of the water pump pipe and the thermostat housing. Apart from the generator and the starter, which are black, the rest was painted "gray enamel." *(Ford Co. ref. 81245-16 and -27)*

Above: Interior access to the starter, engine crankcase and sump. Given the restricted space at the rear, handling, repairing and servicing was complicated. (*Ford Co. ref. 77476-11*)

Above: It was often easier to take the engine out completely rather than do the repairs with it in place in the rear compartment. In an Ordnance workshop somewhere in England, on 15 August 1944, civilian workers get the job done. (*U.S. Army SC-192392*)

Below: A rear view of the Hercules JXD engine and clutch bell. (*Ford Co. ref. 81425-22*)

Timken-Detroit, and the rear tandem was a pair of Ford Model GAKs; all three were of the double reduction type.

Both vehicles were powered by six-cylinder, single-block Hercules JXD petrol engines. The JXD was a 320-ci engine rated at 110 bhp at 3,200 rpm. It was water-cooled from a radiator with gilled tubes and a water pump situated on the left side of the engine block. Two 18-inch diameter, six-blade fans increased the cooling of the 5.87 gallons filling the circuit.

Transferring the power of the engine to the wheels was a Long Manufacturing Co Model 12CB-C single dry-disk clutch running through a Warner Gear Model T95-1MF gear box (four forward and one reverse) and a two-stage transfer box, also by Warner Gear (Model J5), which was also used to engage the front differential when changing from 6×4 to 6×6.

With a Zenith 29W-12 carburetor, a fuel pump and a 54-gallon tank, the vehicle had a range of somewhere between 100 and 250 miles off road and between 200 and 400 miles on the road, or an average consumption of between 3.70 and 7.40 mpg; these data varied depending on the type of road surface, the running condition and, above all, the driver's driving style, as there was no regulator on the engine. It was therefore up to the driver to ensure the engine did not overspeed, particularly in the lower gears; driving had to be done by ear with a light foot on the accelerator. The maximum speeds were 56 mph on road and between 6 and 30 mph off road.

The Electrical Circuit

A traditional design, the electrical circuit was 12 volts, from a battery fitted into the engine compartment and a 50-amp dynamo delivering 750 watts. The current was stabilized by a Ford or Auto-lite regulator.

Although the whole installation was 12 volts, including all the internal and external lighting, there was, however, an exception. The blackout driving lights were only available in 6 volts, so a resistor was inserted into the circuit, inside the instrumental panel.

For the radio installation in the double radio-set configuration to work, there was the issue of fitting an independent electric generator to supply power when the engine was not running. However, there was not enough space for such equipment. Several trials, like installing a second generator and a second regulator, were not immediately satisfactory, even more so as they did not provide a solution when the engine was not running. All the requests made during the trials on the T22E2 pre-production series vehicle were answered, until a real solution could be found, by installing a junction box in each of the side bulges of the body.

The Hydraulics

Be it the M8 or the M20, the hydraulic system had three parts. The control layout set up at the front for the mechanical elements in the rear avoided a complicated set of linkage rods.

1. The braking system normally consisted of a master cylinder under the instrument panel assisted by a Hydrovac servo-brake, placed on the left in the engine compartment. The pressure applied to the brake pedal was thus amplified and transmitted to each wheel. Inside each drum, two slave cylinders made two brake segments move to slow the wheels or stop them if necessary.
2. The clutch control was also hydraulic. As with the brakes, the pedal acted on a master cylinder under the instrument panel which activated the slave cylinder, mounted on the left side of the

Top: A shot of the front axle, the steering arm and the drive shaft between the differential and the transfer box. (*Ford Co. ref. 77476-27*)

Middle: A shot of the middle axle, looking towards the front, and the position of the transfer box. (*Ford Co. ref. 77476-26*)

Right: Fitting the drive shafts was a real feat of engineering where all elements have their own clearances. (*Ford Co. ref. 77476-30*)

Above: A shot of the passage out of the gearbox and the return towards the twin axles from the transfer box. (*Ford Co. ref. 77476-18*)

Above right: With two receiving cylinders on each wheel, the braking system needed small pipes to spread the hydraulic fluid from the main piping. (*Ford Co. ref. 77476-17*)

Below: Despite their designation as "armor," the M8s and M20s were no more complicated to maintain than a 6×6 lorry. Summer 1944, in the middle of the Normandy *bocage*, replacing the front differential of this M8 using the cable from a Dodge's winch and a hoist mounted to its front end. (*U.S. Army Signals Corps*)

The configuration of the engine compartment with all its early elements, including the little sensor for the fire detection system, near the oil can on the left. (*Ford Co. ref. 77476-15*)

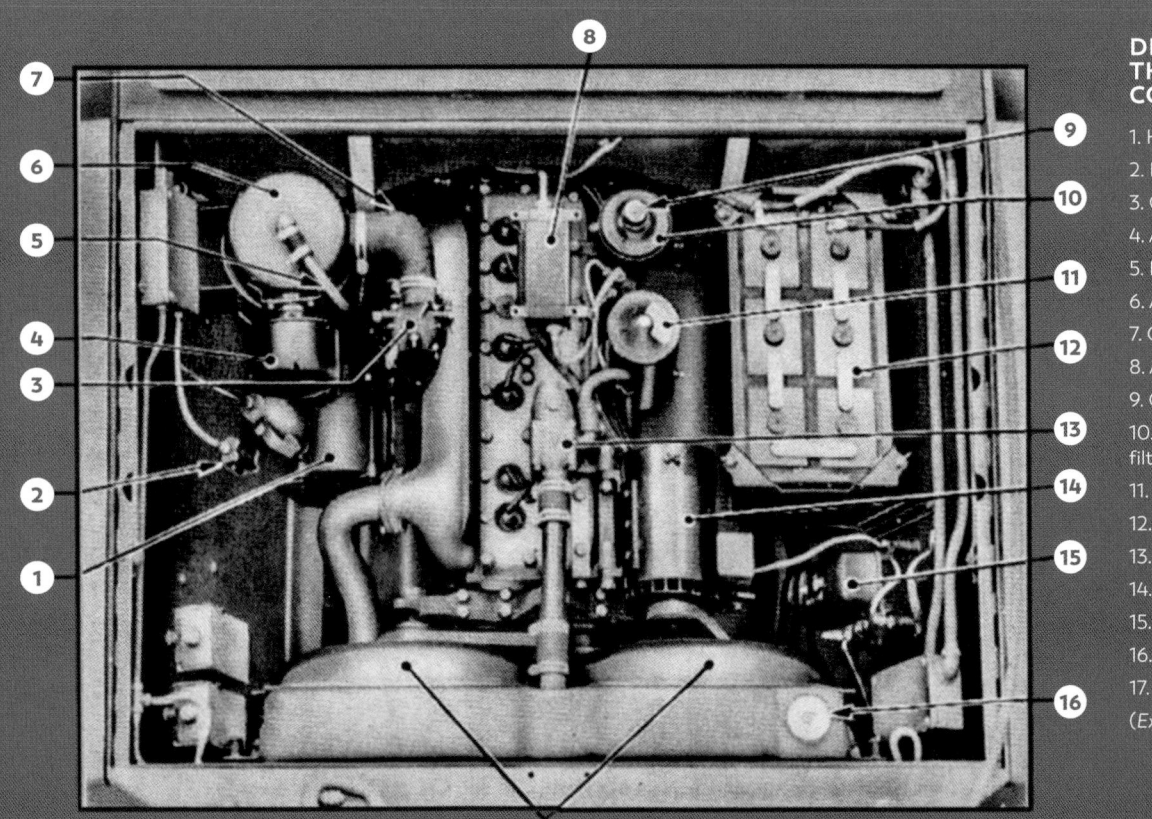

DETAILS OF THE ENGINE COMPARTMENT

1. Hydrovac
2. Fire detection sensor
3. Carburetor
4. Air filter inlet
5. Hydrovac air intake
6. Air filter
7. Carburetor air connection
8. Air intake for the ignition
9. Oil filler
10. Crankcase ventilation filter
11. Oil filter
12. Battery
13. Thermostat
14. Generator
15. Regulator
16. Radiator filler cap
17. Fan fairings

(*Extract of TM 9-743 of 1943*)

gear box, which mechanically transmitted the hydraulic impulses to the clutch.

3. The movement of the accelerator pedal was directly transmitted to a third master cylinder, placed in front of the accelerator, whose hydraulic liquid reserve was situated in the small cylindrical reservoir mounted in front of the co-driver. The action was transmitted to a slave cylinder under the carburetor to which it was linked by a set of short linkage rods.

Paintwork

It was not easy to find factory information about the shades of paint used during manufacture. In technical instructions dated 15 January 1943, however, the Ordnance Department gave clear and detailed guidance.

1. The outside of the vehicles had to be painted olive drab (Enamel, synthetic, Olive Drab)

2. For camouflage reasons, the War Department Registration numbers were to be painted on each side in 2-inch-high characters, using drab blue (Enamel, Blue Drab). They could thus appear on aerial photos but were not visible to the human eye beyond 75 feet.

3. For the engine, which is rarer in Army specifications, high-resistance gray paint (Gray Enamel) was required.

4. The fillers and oil change, and all the other systems for lubrication, had to be painted bright red (Enamel, synthetic, Gloss Red). Each greasing point had to be indicated by a 0.75-inch disc of the same color.

Fire Detection

To warn the driver that there was a fire in the engine compartment, a detector circuit was fitted. It consisted of a sensor on the left side of the body near the Hydrovac, and a red warning light on the left of the dashboard; below that was a test button to check everything was working properly.

By means of a circular dated 6 August 1943, however, it was soon decided to abandon the detection system on the M8 and M20 production lines, since Ford would have to modify the electrical circuits and dashboard for the remainder of the production run. For the vehicles already delivered, the units supplied with them, or the depots, had to get rid of this detection circuit. If this was not possible, for reasons of lack of time or manpower, the fire detection system could simply be disconnected from the circuit and the dashboard light and instruction panel painted over with Olive Drab.

M8 and M20 Evolution

Unlike most vehicles of the time, the M8s and M20s were not majorly modified and/or improved during their production, just updated.

This evolution was clearly evident in the Standard Nomenclature List (SNL-G) for the two vehicles. Brought together mostly in the table below, these modifications give a good idea of the different configurations the M8s and M20s adopted.

Engine Compartment

Apart from a change of cylinder head, the engine block remained the same; the accessories evolved with time. A new head unit, for the shared crankcase ventilation filter and oil filler, was fitted.

Left: Easily visible is the low-mounted exhaust pipe under the right rear fender. This concerned M8s with serial numbers lower than #3908, except the odd numbers between #1361 and #3907. For the M20s, this was applicable up to #858. (*Ordnance ref. 549*)

Below left: Identifying an engine compartment from the end of production is easily done by locating the water circuit thermostat housing. Its cylindrical shape is unmistakable. (*Ordnance ref. 1528*)

Below right: An extract from SNL G-136/176. The first model of exhaust with the pipe terminating in the lower position.

Right: The later fitting of the exhaust pipe exited through the rear fender, which was ribbed to make it more rigid. These modifications concerned all M8s with builder's numbers above 3908 and odd numbers between #1360 and #3907. For the M20s, this change was applied from #859. (*Ordnance ref. 1501*)

Below: An extract from SNL G-136/176. The second model of exhaust, with the pipe terminating in the higher position.

The fuel pump with a glass bowl became a multi-membrane model without a sight glass. The engine temperature thermostat and its square housing, which became cylindrical, also changed. The radiator was given a new filler cap and the Ford generator evolved and was partly included in production alongside an Auto-Lite model; two variants of the regulator were also included. Although not considered an improvement, a new protective cover for the bottom of the engine was fitted; it was entirely interchangeable with that from the early production vehicles.

The mechanical modification which had the greatest effect was most certainly the new position of the exhaust pipe. The original location was thought to be too low and too susceptible to damage, so it was modified to come out higher up, through the rear face of the fender. This change meant a whole new exhaust and new ribbed fender panels. These modifications concerned all vehicles after #3908 and odd numbers between #1361 and #3907. For the M20s, the change took place after #859.

Bassanio, Italy, 29 May 1944. The 27th vehicle of Troop A, 91st Cavalry Reconnaissance Squadron, attached to the Fifth U.S. Army, entering the village. There were big differences in the rear bodywork between early and later models, depending on where the exhaust pipe came out. It is seen here at the bottom of the fender. (*U.S. Army Signal Corps*)

MAIN MODIFICATIONS FOLLOWING THE MAKER'S OWN NUMBERING (MFR. No.)

HEAD, engine cylinder, assembly (High compression)	M8	After #6494 and on all odd numbers above #2143
	M20	After #1597
BREATHER, engine oil filler tube, assembly	M8	After #3000 and on all odd numbers above #1117
	M20	M20 after #450
PUMP, fuel, engine, assembly, 1st model	M8	Before #8694 and on odd numbers between #9490 and #10186
	M20	Before #2505
PUMP, fuel and vacuum, engine, assembly, 2nd model	M8	From #8694 to #9490 and on all vehicles after #10186
	M20	After #2505
EXHAUST PIPE & MUFFLER (new model)	M8	After #3908 and all odd numbers between #1361 and #3907
	M20	After #859
RADIATOR FITTING	M8	After #8934
	M20	After #2300
GENERATOR FORD, 1st model D67368	M8	Before #502 and before #497
	M20	Before #30
GENERATOR FORD, 2nd model D67368-A	M8	From #502 to #9490 (even numbers) and from #497 to #9489 (odd numbers)
	M20	From #31 to #2212
GENERATOR AUTOLITE	M8	From #8283 to #9490 and on the vehicles after #9978
	M20	After #2213
REGULATOR, assembly, 1st model C118349	M8	From #1 to #779 and odd numbers from #781 to #999
	M20	From #1 to #115
REGULATOR, assembly, 2nd model C118079	M8	From #780 to #998 and all numbers above #1000
	M20	After #115
GAUGE WATER TEMPERATURE, 1st model A283078	M8	From #1 to #8113 (odd numbers) and from #2 to #7836 (even numbers)
	M20	Before #2146
GAUGE WATER TEMPERATURE, 2nd model B249030	M8	Above #8113 and above #7836
	M20	After #2146
GAUGE OIL PRESSURE, 1st model B246994 (80 lbs)	M8	From #1 to #8113 (odd numbers) and from #2 to #7836 (even numbers)
	M20	Before #2146
GAUGE OIL PRESSURE, 2nd model B249030 (50 lbs)	M8	Above #8113 (odd numbers) and above #7836 (even numbers)
	M20	After #2146 up to #3320
GAUGE OIL PRESSURE, 3rd model B210455 (120 lbs)	M8	Above #9005 (odd numbers) and above #10987 (even numbers)
	M20	After #3320
GAUGE FUEL, 1st model B199960	M8	From #1 to #8113 (odd numbers) and from #2 to #7836 (even numbers)
	M20	Before #2146
GAUGE FUEL, 2nd model B209888	M8	Above #8113 and above #7836
	M20	After #2146
SPRING, 11-leaf, front assembly, 1st model D67331	M8	Before #896 and before #681
	M20	Before #48
SPRING, 11-leaf, front assembly, 2nd model D67450	M8	Above #896 and above #681 (replaced by the 13-leaf)
	M20	Above #48
SPRING, 13-leaf, front assembly, model D67515	M8	From #683 to #7467 (odd numbers) and from #898 to 37468 (even numbers)
	M20	Above #48
SPRING, rear, assembly, 1st model D67332	M8	From #1 to #897 (odd numbers) and from #2 to #666 (even numbers)
	M20	Before #238
SPRING, rear, assembly, 2nd model D67451	M8	Above #897 and from #666
	M20	After #238
TURRET TRAVERSING MECHANISM, 1st type	M8	Before #1025 and before #2686
TURRET TRAVERSING MECHANISM, 2nd type	M8	After #1025 and after #2686
COMPASS (co-driver's)	M8	Installed after #1000 and after #843
	M20	Installed after #3411
WINDSHIELD & WIPER		Installed on demand
SPONSON CLOSED		Exterior sponsons with panels. Installed on demand.

FIRST GENERATION

SECOND GENERATION

THIRD GENERATION

The Suspension

The intensive trials with the 159 early production M8 Light Armored Cars had a direct effect on the front suspension.

The first examples of sets of 11-leaf springs were modified and reinforced; it was only later that the 13-leaf sets appeared. These better-performing sets were applied to the M8s from #683 to #7467 (odd numbers), and #898 to #7468 (even numbers), and then all the vehicles built after these two series. For the M20s, the 13-leaf sets of front springs were applied to all vehicles after #48.

Replacing the sets of 11-leaf springs with 13-leaf sets, which were longer, was done by reversing the spring shackle.

The Instrument Panel

Three models followed each other on the M8s and M20s. The early version was not codified because it was under development; the voltmeter and fire detector were subsequently removed.

THE INSTRUMENT PANEL

1. Headlight controls
2. Voltmeter
3. Instrument panel lights
4. Circuit breaker for the siren and voltmeter if present
5. Lighting circuit breaker
6. Circuit breaker for the instruments and fire detector when present.
7. Ammeter
8. Blackout lights control
9. Instrument panel lighting rheostat
10. Engine temperature
11. Siren
12. Fuel gauge
13. General switch
14. Starter
15. Speedometer
16. Trip counter re-set
17. Oil pressure
18. Fire detection test button
19. Warning light for the fire detector in the engine compartment

The second model was fitted in vehicles with odd serial numbers up to, but not including, 8113 and below 7836 for Saint Paul even-numbered machines. The third model of instrument panel was fitted to all vehicles above these two numbers. In the M20s, the first panel was used until #2146 and the second design was fitted to the remainder. Changes were also made to control switches which were no longer pulled but turned.

The Turret Mechanism

A single-speed manual swiveling mechanism was fitted on the early models, before #1025 and #2686 for the Chicago and Saint Paul factories respectively.

Subsequently, from #1025 and #2686 onwards, the swiveling mechanism was fitted with a lock and had two speeds, controlled by a push button in the center of the handle. A simple push on the button was enough to move to the faster speed. The turret swiveled to the right when the wheel was turned clockwise and to the left when turned anticlockwise.

The Compass

Delivered late by its manufacturer, and with two types of quite different mountings, the Sherrill AEG model was only fitted to M8s after #843 and #1000. For the M20s, it was only fitted after #3411.

Above left: M8 #2940 and its first-generation dashboard. (*Ordnance ref. 686*)

Above right and below: M8 #8325 and its third-generation instrument panel. To make handling easier and improve instrument panel visibility despite the size of the steering wheel, all controls were identified. (*Ordnance ref. 1504 and Ford Co. ref. 81425-8*)

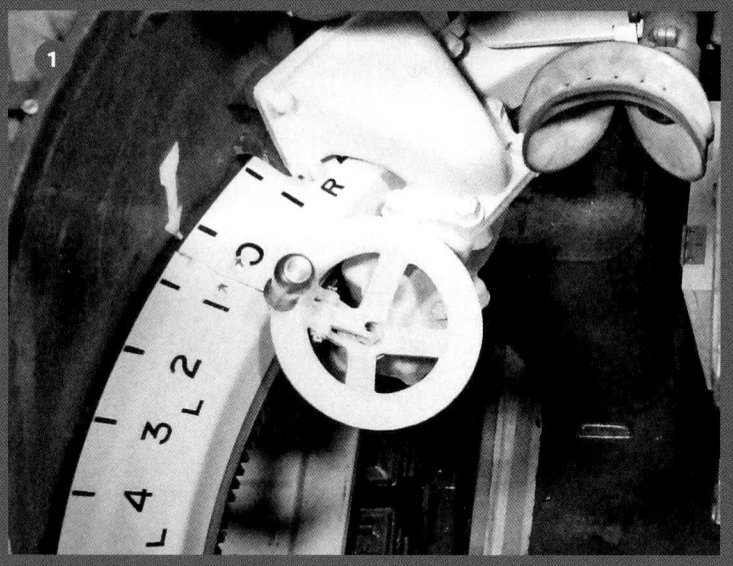

THE TURRET SWIVELING MECHANISM

1. The turret swiveling mechanism on the first M8s had a single speed and was in direct drive on the ring gear. It was installed on Chicago vehicles before #1025 and before Saint Paul's #2686. (*Ford Co. ref. 77441-24*)

2. The second version of the swiveling mechanism had two speeds and a lock. The indicator for where the gun was pointing in relation to the vehicle's direction of travel was a simple arrow painted on the turret. (*Ordnance ref. 1510*)

3. An exploded view of the second swiveling mechanism model. (*Ford Co. ref. 81425-17*)

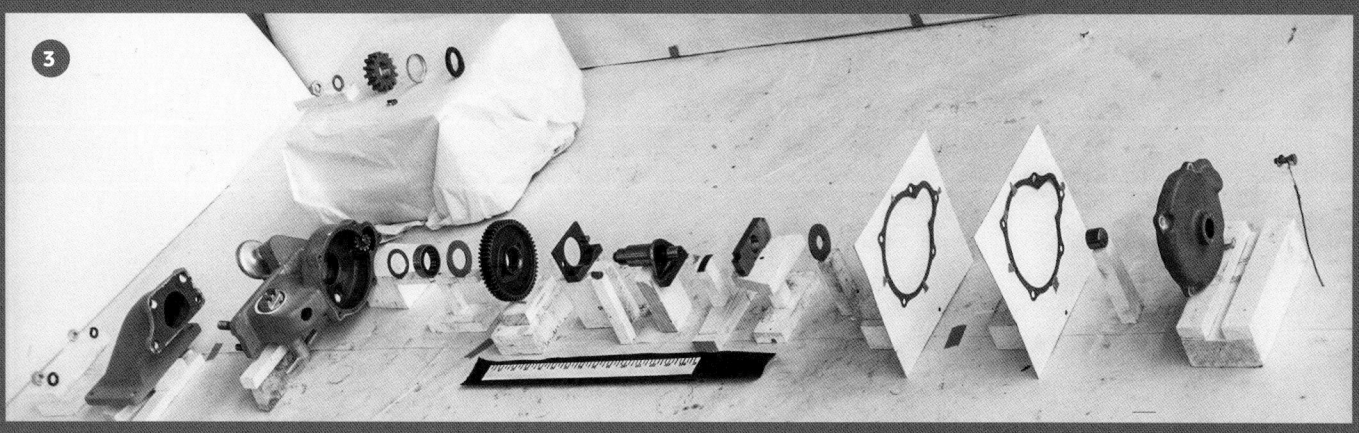

Meanwhile, the smaller M4 Hull Compass was used by the car commander.

The Windshield

Installing protective panes on the two openings of the drivers' positions was an option to improve the crew's comfort. This kit was made up of two paned frames, fitted in place of the two armored panels, each having an arm and windscreen wiper linked by rods to a small vacuum-powered motor. It was necessary, therefore, to install flexible piping between the engine compartment and the front of the vehicle. The irony was that the kits did not include this vital piping; Ford only supplied rolls of 100 feet, enough to equip three or four vehicles. A metal locker was fitted to the forward hull to stow the windshields.

The Side Lockers

Constantly fitting new equipment to the vehicles made the body layout even more difficult. To try to make up for the lack of space, external locker

THE COMPASS

Below left: The drivers' positions of M20 #3038. The Sherrill compass is mounted in a first model mounting, with four attachment points, in front of the co-driver; the instrument panel is third generation and the vacuum motor for the windscreen wipers has been fitted complete with piping. (*Ordnance ref. 2803*)

Below right: Extract from TM 9-743 of 1943; this is the small navigation M4 Hull Compass, which was used while waiting for the Sherrill AEG Models.

kits were made available. They were fitted to each flank, in place of the mine racks. Made up of three sides, they had no back and, like the lid, were attached by several spot welds. On the initial versions, they made it impossible for the crew to climb into the vehicle from the side; it was only the final version that featured a notch that could be used as a step.

THE GLAZED WINDSCREENS

Left: What is interesting about this photo is the date—23 November 1943 at the Aberdeen Proving Ground—indicating the trials for glazed protection of the drivers' posts was not a later development but rather from mid-production. (*Ordnance*)

Right: Whether the armored panels of the drivers' posts were open or closed, the glazed windshields could be installed easily. Their stowage locker was mounted on the front where it has been painted with a white star. (*Ordnance*)

Below: A parade in the streets of Oslo, Norway, for Allied Forces Day in June 1945. The M8 Light Armored Cars of the 474th Infantry Regiment are all fitted with the windshield kit.

THE EXTERNAL STORAGE BOXES

Almost the silhouette's final development with storage boxes fitted on the sides between the two differentials, and the box on the front for the windshields. (*Ordnance ref. 1503*)

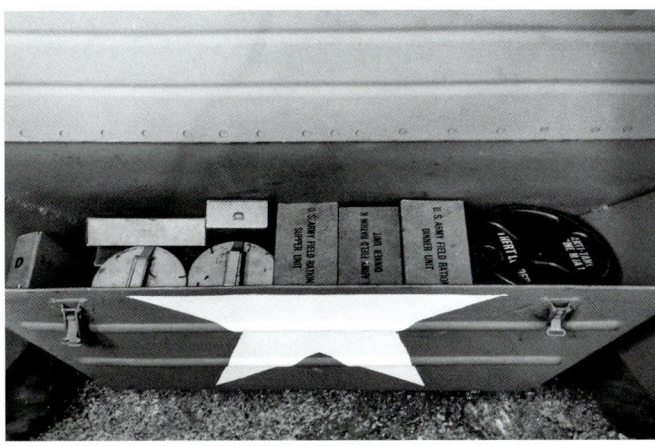

Above and above right: The assessment of the lateral storage boxes started in September 1944, but they were not available until much later. Simply attached by several spot welds, they were not really part of the overall structure of the vehicle. (*Ordnance ref. 1515*)

Below: The definitive version of the side storage box was the one with the notch used as a step. One can appreciate the importance of the

lifting and maintenance points at the four corners of the vehicle with this one being loaded in a U.S. port. (*U.S. Army Signal Corps*)

Below right: Various small details were also modified towards the end of the production: the siren was replaced by a horn; the towing shackles were replaced by hooks which were easier to use since they did not need to be undone. (*Ford Co. ref. 81425-17*)

The Radios

Below: The right-hand flank of an M20 and its upper right-hand locker occupied by the armored force's SCR-508 radio set. (*Ordnance ref. 2807*)

Bottom left: On the left side of the same vehicle, a medium-range SCR-506. (*Ordnance ref. 2808*)

Bottom right: All these radio sets were standard models. There was no special design for the M8 or M20. This SCR-610, for example, works in the same way as if it were installed in a Jeep. (*Extract U.S. Army SC-257532*)

As reconnaissance, specialized service, or command vehicles, the M8s and M20s needed to be able to communicate quickly to a wide range of recipients.

During the whole project, numerous tests were run to determine which had the best performance. In May 1943, a dozen or so sets were tried out and six were selected: the SCR-506, SCR-508, SCR-510, SCR-528, SCR-608, and SCR-610. These sets were mostly short-range (less than 25 miles) and were theoretically installed as follows:

- in the upper part of the body's left bulge, the SCR-506 or the SCR-508 were considered the primary means of communication.

- in the upper part of the body's right bulge, the SCR-510, SCR-608, SCR-528, or SCR-610.

Whether one set or another was installed or not depended on the unit and its mission. Artillery radios were therefore fitted into the armored cars of Tank Destroyer units.

The SCR-506

Used by the Armored Force, the Tank Destroyers, and the Cavalry, the SCR-506 was a medium-range set (up to roughly one hundred miles) made up, apart from its accessories, of a BC-652 receiver and a BC-653 transmitter. Operating it required a 15-foot aerial.

It communicated mainly with the SCR-131, SCR-171, SCR-187, SCR-193, SCR-203, SCR-209, SCR-210, SCR-238, SCR-245, SCR-259, and SCR-287. This type of equipment was only used at the end of 1943/beginning of 1944.

The SCR-508

Mainly used by the Armored Force in the company commander's vehicle, the SCR-508 was made up of two BC-603-Cs for receiving and a BC-604-C for transmitting, with a range of between 5 and 20 miles.

It needed a 9-foot aerial on an FT-237 Mount to work. It communicated mainly with the SCR-293, SCR-294, SCR-508, SCR-509, SCR-510, SCR-528, and SCR-538.

The SCR-510

Used mainly by the Armored Force, the SCR-510 consisted of a BC-620 for receiving and transmitting short range (5 miles). It needed an 8-foot aerial. It communicated mainly with the SCR-293, SCR-294, SCR-508, SCR-509, SCR-510, SCR-528, and SCR-538.

The SCR-528

Another Armored Force set, used within the battalion-platoon radio net, the SCR-528 consisted of a BC-603-C and a BC-604-C transmitter, with a range of 7 miles. It needed a 9-foot aerial on an FT-237 Mount. It communicated mainly with the SCR-293, SCR-294, SCR-508, SCR-509, SCR-510, SCR-528, and SCR-538.

The SCR-608

The artillery version of the SCR-508 was usually made up of two BC-683-(A)s for receiving and a BC-684 for transmitting, with a range of between 5 and 50 miles. It also needed a 9-foot aerial on an FT-237 Mount. It communicated mainly with the SCR-608, SCR-609, SCR-610, and SCR-628.

The SCR-610

Mainly used for artillery spotting, the SCR-610 was composed mainly of a BC-659 transmitter with a range of between 3 and 5 miles. It needed an 8-foot aerial to communicate effectively, mainly with the SCR-608, SCR-609, SCR-610, and SCR-628.

The Electrical Kits

It was quickly noted the vehicle's electrical installation might not be enough for using two radio sets despite a

Top: The SCR-608 (two receivers, one transmitter) was for artillery use. (*TM 11-620*)

Middle: Trial fitting of the second generator on the left side of the engine compartment. As a precaution, the exhaust pipe at the manifold was fitted with a heat deflector. (*Ford Co. ref. 81425-3*)

Right: The layout of the second generator in the rear left-hand section of the engine compartment of an M20. (*Ordnance ref. 2811*)

12-volt system and a 55-amp high-output generator. Various solutions were trialed unsuccessfully, like fitting a small independent generator or increasing the voltage to 24 volts. The problem raised several observations.

1. When stopped, the vehicles did not need to transmit frequently, so the battery installed in the engine compartment could handle brief exchanges.

2. During operations, the M8s were, in theory, never at a halt and, even with the engine just ticking over, the current produced was easily sufficient for signals. On the move, using both sets at the same time did not raise any problems.

3. The M20s, being service vehicles, had standby periods that could be much longer; the engine ticking over did not produce enough current for intensive use of both radio sets.

4. The requirement for M20s fitted with two radio sets (not more than 25 percent of available vehicles) was low.

In July 1943, Ford, working together with the Signal Corps Laboratories, was seeking an alternative solution to produce more energy without fundamentally modifying the mechanicals or re-equipping the assembly lines. It took them several months to find a solution: a kit for installing a second generator. Placed on the rear left-hand side of the engine compartment, it was driven, like the original generator, by a belt powered by the engine and installed on the free groove of the pulley on the left ventilator fan's axis. The kit had all the electrical connections, all the accessories, and the main bracket to fix it to the side of the engine compartment.

In February 1944, the first kits were ready to be tested and were sent to the Lima Tank Depot for installation in five M20s. In March, it was confirmed that only the M20s to be fitted with two radio sets would be given a second generator.

The kit would be installed by the workshops of three specific depots because only they knew where the vehicles were going and what their radio needs were; no M8s were fitted with a second generator.

In March, 500 kits were ordered under the reference Kit B248984; four were distributed for tests, of which

Top left: A view of the fixtures for the second generator and the new tabs welded to the body. (*Ford Co. ref. 81426*)

Middle left: Nothing could prevent regulations from being breached and other methods and means of communication from being used. (*U.S. Army Signal Corps*)

Left: The two M8 intercom casings were situated to the right of the co-driver; the rest of the apparatus was in the fighting compartment. (*Ordnance ref. 1506*)

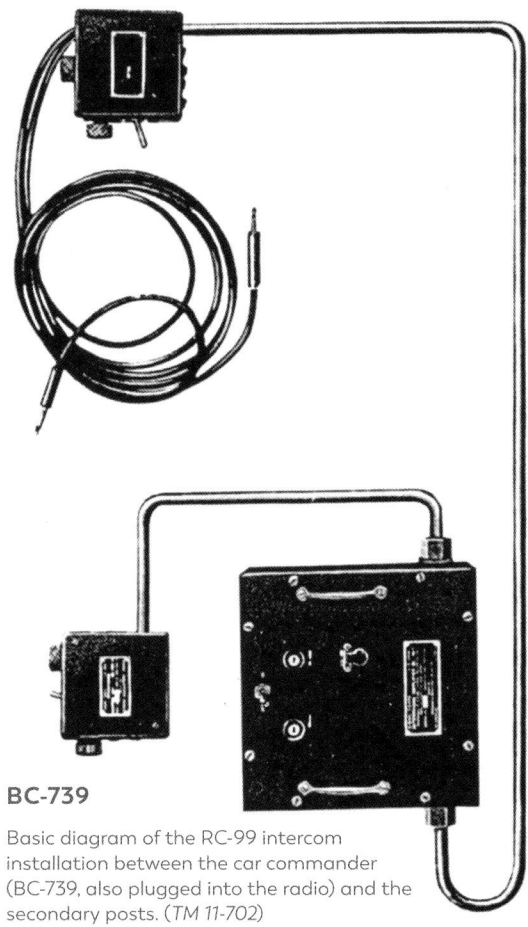

BC-739

Basic diagram of the RC-99 intercom installation between the car commander (BC-739, also plugged into the radio) and the secondary posts. (*TM 11-702*)

one went to the Tank Destroyer Board, the other 496 went to the Depot Modification Sections (175 examples to the Lima Tank Depot, 200 to the Chester Tank Depot, and the remaining 121 to the Richmond Tank Depot). These three centers were entrusted with modifying the M20s; this concerned vehicles built before #1512. For the remainder, supplies were ordered as required.

The Intercom

The date on which the intercom was introduced into the study program for the M8 and the M20 is not known, but 29 July 1943 was a turning point; the installation in the M8 was made official and it was confirmed it would not be fitted in the M20s. This installation, designated the RC-99 in its 12-volt version, was made up of a BC-367 amplifier, a BC-739 Commander's Control Box and two BC-606 Driver's Control Boxes. The HS-30 Headsets and T-30-A microphones were plugged into each control box. This equipment enabled the car

commander in the turret to communicate with the two men in the drivers' positions.

Field Phone

It may seem strange to talk about field phones in an armored vehicle but, in February 1944, the Field Artillery Board asked whether it was possible to install a Reel Unit with two Type DR-4 telephone cable reels on the rear of the M20s. The Ordnance Department replied favorably to the request, had two standard RL-31s modified, and designed an attachment system. On 7 March, the two kits were delivered to the Army Ground Forces Headquarters Test Center. Major functioning defects were soon noted and the project suspended, though not canceled.

Top: This SCR-528 is installed behind the M8 driver. (*Ordnance ref. 1507*)

Above: An SCR-506 has been fitted behind the co-driver in the bulge on the right side of the body. (*Ordnance ref. 1508*)

The Deep Water Fording Kit

Enabling vehicles to move a certain distance through shallow water to the beach during an amphibious operation was the aim of this waterproofing process.

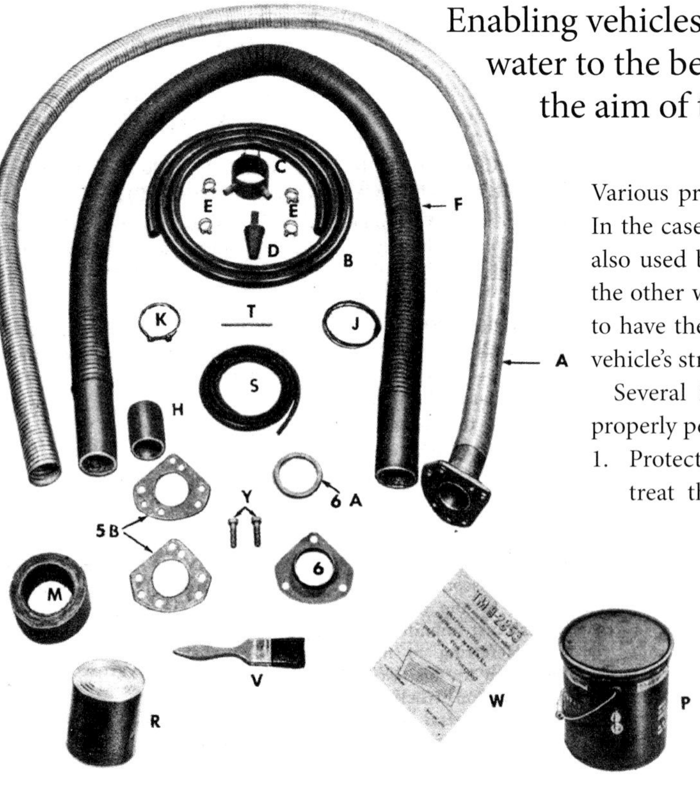

Various preparation kits were available from the Ordnance Department. In the case of the M8s and M20s, this was reference G-9-5700771; it was also used by the White, Autocar and Diamond T Half-Tracks. Unlike all the other wheeled vehicles, the half-tracks and armored cars did not need to have their fuel tanks modified because they were fitted high up in the vehicle's structure.

Several important steps were to be carried out when waterproofing properly per the handbook:

1. Protecting the weapons: on the M8 Light Armored Car, clean the gun, treat the pointing/swiveling mechanisms and the breech with type

Above: Waterproofing School, Codford, England, 1 April 1944. The crews are learning how to master waterproofing techniques and how to drive partly submerged. This system increased the fording capacity from 25 inches in normal conditions to more than 39 inches. (*U.S. Army SC-203709*)

Left: Kit G-9-5700771 in detail with its exhaust pipe (A), its carburetor air-intake pipe (F) and the various connection and waterproofing items, including the pots of special grease. (*TM 9-2853*)

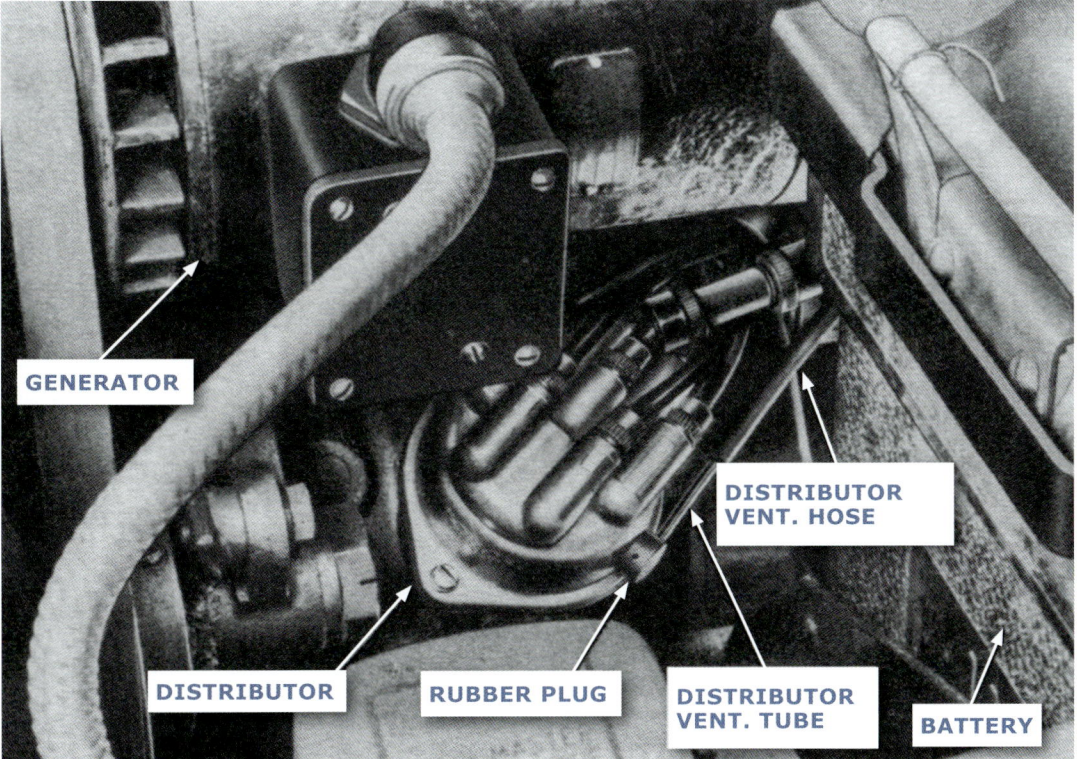

GENERATOR

DISTRIBUTOR
VENT. HOSE

DISTRIBUTOR

RUBBER PLUG

DISTRIBUTOR
VENT. TUBE

BATTERY

A vital step was the one that ventilated the carburetor head properly so the ignition cycle was not affected. This operation was done by installing small rigid pipes. (*TM-9-2853*)

AXS674 protective oil, cover the muzzle of the cannon with two pieces of crossed adhesive tape, roll a third round the cannon, securing the ends of the two pieces of sticky tape already in place. In extreme conditions, this cap could be removed by shooting shells without explosive heads, like grapeshot or smoke shells, before using explosive ammunition.

2. Protecting the electrics: all the electrical parts that were likely to be damaged by water had to be protected by adhesive tape and/or covered in special anti-dispersion grease.

3. Protecting the mechanicals: all mechanical parts of the transmission and/or moving parts must be copiously covered with protective grease.

4. Protecting the drivers' posts: cover all the little openings with adhesive tape and seal off the steering column opening with asbestos grease.

5. Protecting the engine: the basic points of this preparation were undoing the exhaust pipe at the manifold and fitting a second flexible pipe to extend out in the open air on the rear right-hand side; fitting a similar system for the carburetor air intake on the side of the turret by removing the air-filter joint; ventilating the crankcase via the oil filler and the distributor; and placing a wooden chock to prop the left-hand engine cover half open.

The vehicle thus prepared could only be used, at most, for eight minutes; after being used during this time, the vehicle had to be immobilized and the reverse process carried out.

Below: Connecting the extension of the air intakes with the help of hoses.

1. Air-intake hose ventilating clamp

2. Carburetor air-intake hose

3. Crankcase ventilating hose

4. Hydrovac air-intake pipe.

5. Carburetor air-intake elbow

6. Distributor ventilating hose

(*TM 9-2853*)

HYDROVAC
VENTILATING
HOSE

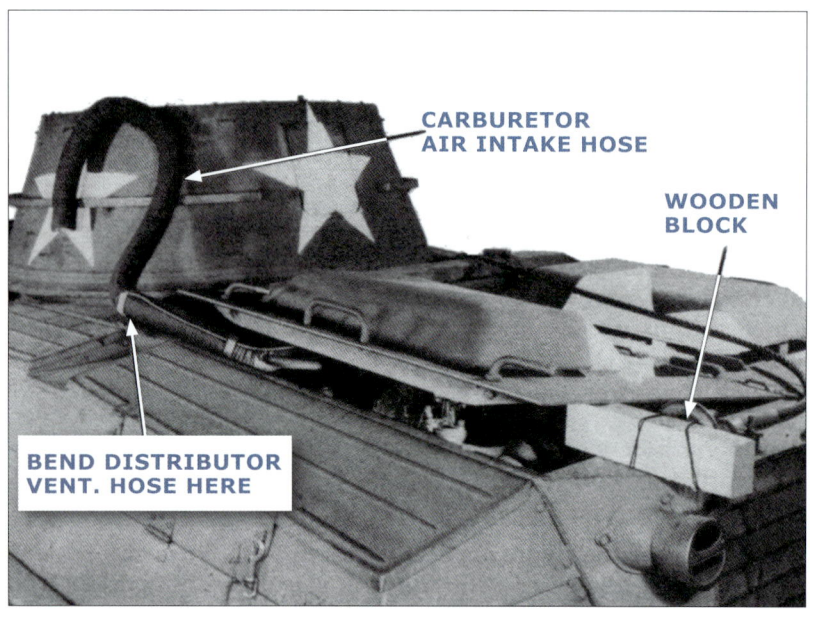

CARBURETOR
AIR INTAKE HOSE

WOODEN
BLOCK

BEND DISTRIBUTOR
VENT. HOSE HERE

FLEXIBLE METAL
EXHAUST PIPE

WOODEN SUPPORT
EXHAUST PIPE

TAIL LIGHTS
SEALED WITH
ASBESTOS GREASE

CARBURETOR
AIR INTAKE
HOSE

WOODEN
BLOCK

CONNECT UNIVERSAL
EXHAUST PIPE
FLANGE HERE

TAIL LIGHTS
SEALED WITH
ASBESTOS GREASE

Driving was also heavily restricted: before the vehicle went into the water, the engine had to be preheated, but not up to its maximum temperature (so as to avoid thermal shock); the tires had to be under-inflated—50 percent of what was normal; the transfer box was put into 6×6 and the gearbox into first gear with reduction engaged. Anticipating the front differential entering the water and its contact with the soft bottom, the accelerator had to be pressed down three-quarters of the way and not released before the vehicle was out of the water, keeping up a maximum speed of 3–4 mph so as not to create too much of a bow wave. Once out of the water, and as soon as operations permitted, without, nonetheless, going beyond the regulation time period in this configuration (thereby risking irreversible damage), all the fording kit elements had to be taken off and inspected at length to ensure there was no water ingress, the brakes cleaned, the tires re-inflated, the oil filter re-fitted, oil refilled, the vehicle completely rinsed with fresh water and, finally, all requisite points greased.

Workshops with pools were opened before the landings in Normandy to instruct the drivers but putting all this into practice was very restrictive and relatively few wheeled vehicles underwent the entire process. In a lot of cases, the vehicles were only partly waterproofed because, in the end, few of them were in the first wave of the landings. Only the carburetors and exhaust pipes were protected so the dismantling work needed was limited and, in a lot of situations, everything was done to avoid the vehicles being used in water.

Above: The positioning of the air-intake hoses for the carburetor, the distributor venting hoses, and the wooden block to keep the engine panel open. (*TM 9-2853*)

Left: Position of the flexible exhaust pipe that had to replace the entire exhaust pipe system, including the muffler. It ended high up, attached to a wooden support. The rear lights were sealed with asbestos grease. (*TM 9-2853*)

5

M8 and M20 in Europe

Left: On the Champs Elysées in Paris, 29 August 1944. M8s of the 28th Cavalry Reconnaissance Troop of the 28th Infantry Division on parade following the city's liberation. (*U.S. Army SC-193390*)

Primary Allocations in Europe

With regard to reconnaissance units, at divisional level or not, there was almost no change in allocations between May 1943 and the end of the war.

Below: Schwäbisch-Gmund, Germany, 31 miles east of Stuttgart, 4 July 1945. A celebration day for these GIs. The 92nd Cavalry Reconnaissance Squadron of the 12th Armored Division parades past Major General H. H. Morris. (*U.S. Army SC-208821*)

The organization of the small units was based on platoons, gathered in threes or fours, each receiving three M8s, with exceptions like the Tank Destroyer Battalions or the 1st, 2nd, and 3rd Armored Divisions' (divisions considered as heavy according to 1942 organizational table) reconnaissance units.

The Armored Divisions

For so-called "light" armored divisions, according to the September 1943 organizational table, the allocation was for 53 M8 Light Armored Cars and one M20 Armored Utility Car, almost all gathered in a cavalry reconnaissance squadron (regiment) which numbered four troops of three platoons (see table below).

The Infantry Divisions

The cavalry reconnaissance troop of an infantry division was organized like that of an armored division.

The only difference was that it had three platoons instead of four, for a total of 39 M8s. In the North-West European Theater of Operations (ETO), no fewer than 42 infantry divisions were engaged between June 1944 and May 1945. The reconnaissance units incorporated into the infantry divisions bore the division's number. For example, the 5th Infantry Division had the 5th Cavalry Reconnaissance Troop.

M8/M20 ALLOCATIONS		
1 Armored Division (light)		
	53 × M8	**1 × M20**
Headquarters Company—Armored Section	2	—
Headquarters Division Artillery	—	1
Cavalry Reconnaissance Squadron		
Cavalry Reconnaissance Squadron		
Headquarters—Communication Section	3	—
Cavalry Reconnaissance Troop Headquarters (×4)		
Headquarters Section	2 (×4)	—
Maintenance Section	1 (×4)	—
Reconnaissance Platoon (A)	3 (×4)	—
Reconnaissance Platoon (B)	3 (×4)	—
Reconnaissance Platoon (C)	3 (×4)	—
2 Infantry Division		
	39 × M8	
Cavalry Reconnaissance Troop		
Cavalry Reconnaissance Troop Headquarters (×3)		
Headquarters Section	3 (×3)	—
Maintenance Section	1 (×3)	—
Reconnaissance Platoon (A)	3 (×3)	—
Reconnaissance Platoon (B)	3 (×3)	—
Reconnaissance Platoon (C)	3 (×3)	—
3 Cavalry Reconnaissance Squadrons (general reserve)		
	36 ×M8	
Cavalry Reconnaissance Troop Headquarters (×3)		
Headquarters Section	2 (×3)	—
Maintenance Section	1 (×3)	—
Reconnaissance Platoon (A)	3 (×3)	—
Reconnaissance Platoon (B)	3 (×3)	—
Reconnaissance Platoon (C)	3 (×3)	—

RECONNAISSANCE UNITS OF THE ARMORED DIVISIONS	
1st Armored Division	81st ARB/CRSM
2nd Armored Division	82nd ARB
3rd Armored Division	83rd ARB
4th Armored Division	25th CRSM
5th Armored Division	85th CRSM
6th Armored Division	86th CRSM
7th Armored Division	87th CRSM
8th Armored Division	88th CRSM
9th Armored Division	89th CRSM
10th Armored Division	90th CRSM
11th Armored Division	41st CRSM
12th Armored Division	92nd CRSM
13th Armored Division	93rd CRSM
14th Armored Division	94th CRSM
16th Armored Division	23rd CRSM
20th Armored Division	33rd CRSM
(ARB = Armored Reconnaissance Battalion, CRSM = Cavalry Reconnaissance Squadron, Mechanized)	

The Independent Cavalry Squadrons

These "separate" units operated at Army or Army Corps level.

COMPOSITION OF CAVALRY GROUPS (ETO)	
2nd Cavalry Group	2nd + 42nd CRSM
3rd Cavalry Group	3rd + 43rd CRSM
4th Cavalry Group	4th + 24th CRSM
6th Cavalry Group	6th + 28th CRSM
11th Cavalry Group	36th + 44th CRSM
14th Cavalry Group	18th + 32nd CRSM
15th Cavalry Group	15th + 17th CRSM
16th Cavalry Group	16th + 19th CRSM
101st Cavalry Group	101st + 116th CRSM
102nd Cavalry Group	38th + 102nd CRSM
106th Cavalry Group	106th + 121st CRSM
113th Cavalry Group	113th + 125th CRSM
115th Cavalry Group	104th + 107th CRSM
(CRSM = Cavalry Reconnaissance Squadron, Mechanized)	

Their make-up was close to that of the armored division's squadron but with three reconnaissance troops instead of four, for a total allocation of 36 M8 armored cars. Twenty-six cavalry reconnaissance squadrons were deployed to the ETO and were gathered in pairs within the cavalry groups. Two independent squadrons operated by themselves, the 91st and the 117th Reconnaissance Squadrons, the former with the Fifth U.S. Army and the latter with the Seventh U.S. Army.

The Tank Destroyer Battalions (Self-Propelled and Towed)

The organization of the Tank Destroyer Command units was special; the reconnaissance component was important for finding enemy armor in advance and selecting the best places to set up anti-tank gun batteries.

A Ford M8 of Troop B of the 24th Cavalry Squadron (4th Cavalry Group) going down a large street towards the Place Ducale, Charleville-Mézières, on 3 September 1944. (*Private Collection*)

ALLOCATIONS IN TANK DESTROYER BATTALIONS (SELF-PROPELLED)		
	M8	M20
Headquarters & Headquarters Company		
Intelligence section	—	3
Reconnaissance Company		
Company HQ, HQ Section	—	2
Pioneer Platoon	—	1
Reconnaissance Platoon (×3)	6	—
Tank Destroyer Company (×3)		
Company HQ, HQ Section	—	6
Tank Destroyer Platoon (×3)	—	18
Total armored cars	6	30

Left: La Spezia, Northern Italy, 24 April 1945. The 3rd Platoon, 92nd Cavalry Reconnaissance Troop, 92nd Infantry Division, entering the town under the gaze of a less than exuberant crowd. (*U.S. Army Signal Corps*)

Below: Passing in front of a wall covered with slogans threatening traitors and those spreading defeatist rumors with death, an unidentified troop goes through the town of Remagen, three days after the discovery that the bridge was intact. (*U.S. Army SC-202343*)

There were two types of tank destroyer battalions:

THE TANK DESTROYER BATTALIONS IN EUROPE, 1944–45
1 Self-propelled[1]
601st TD Bn. (M10/M18)[3]
602nd TD Bn. (M18)
603rd TD Bn. (M18)
605th TD Bn. (M36)
607th TD Bn. (M36)
609th TD Bn. (M18)
610th TD Bn. (M36)
612th TD Bn. (M36)
628th TD Bn. (M10/M36)
629th TD Bn. (M10)
630th TD Bn. (M36)
631st TD Bn. (M10)
633rd TD Bn. (M18)
634th TD Bn. (M10)
635th TD Bn. (M10)
636th TD Bn. (M10)
638th TD Bn. (M18)
643rd TD Bn. (M18)
644th TD Bn. (M10)
645th TD Bn. (M10/M36)
654th TD Bn. (M10/M36)
656th TD Bn. (M18/M36)
661st TD Bn. (M18)
691st TD Bn. (M36)
692nd TD Bn. (M10)
701st TD Bn. (M10)[3]
702nd TD Bn. (M10/M36)
703rd TD Bn. (M18)
704th TD Bn. (M10/M36)
705th TD Bn. (M18)
771st TD Bn. (M10/M36)
773rd TD Bn. (M10/M36)
774th TD Bn. (M36)
776th TD Bn. (M10/M36)
801st TD Bn. (M18)
802nd TD Bn. (M36)
803rd TD Bn. (M10/M36)
804th TD Bn. (M10)[3]
808th TD Bn. (M36)
809th TD Bn. (M18/M36)
811th TD Bn. (M18)
813th TD Bn. (M10/M36)
814th TD Bn. (M10/M36)
817th TD Bn. (M18)
818th TD Bn. (M10/M36)
820th TD Bn. (M18)
821st TD Bn. (M10)
822nd TD Bn. (M18)
823rd TD Bn. (M10)
824th TD Bn. (M18)
825th TD Bn. (M10)
827th TD Bn. (M18)
893rd TD Bn. (M10)
894th TD Bn. (M10)[3]
899th TD Bn. (M10/M36)
2 Towed[2]
614th TD Bn.
648th TD Bn.
679th TD Bn.
772nd TD Bn.
805th TD Bn.[3]
807th TD Bn.
1. Battalions armed with M10, M18 and M36 self-propelled anti-tank guns.
2. Battalions armed with towed 3-inch anti-tank guns.
3. Deployed in Italy only.

1. The self-propelled battalions were equipped in 1944–45 with cannons mounted on M10 Wolverine, M18 Hellcat or M36 Jackson tracked armored vehicles. Only six M8s were present in the reconnaissance company and 30 M20s were shared between the reconnaissance company and the three tank destroyer companies.
2. The towed battalions were equipped with 3-inch guns towed by trucks. There were four M8s in the reconnaissance platoons and 10 M20s.

The divisions sent to the Pacific and Asia went with with their reconnaissance unit, as here with the 40th Cavalry Reconnaissance Troop, 40th Infantry Division, on the island of Luzon, in the Philippines. (*Private Collection*)

In the Field

The first major Allied operations involving the Americans, like those in North Africa and Sicily, progressed without any M8s and M20s participating, their late production preventing them from being engaged before the invasion of Italy. On this front, a large part of the American units was deployed for the landings in the south of France in August 1944.

Only the 34th, 85th, 88th, 91st and 92nd Infantry Divisions, one armored division (1st Armored

Above: Burma, near the town of Lashio, 24 March 1945. PFCs Regis F. Palko and Thomas J. Thompson have just recovered a Japanese horse-drawn cart which they have hooked up to their M8. (*U.S. Army SC-317789-90*)

Left: Advancing cautiously to penetrate the Netherlands, in the commune of Maarland near Eijsden, in Limburg, on 16 September 1944. (*U.S. Army SC-194769*)

Left: Massarosa was the first Italian village to be freed by Brazilian forces alone; the first M8 is stormed by the jubilant population on 17 September 1944. (*U.S. Army SC-194769*)

Below: Units for keeping order in Occupied Germany (the Constabulary Force) used a great number of M8s and M20s to carry out their missions. On 29 March 1949, the 16th Constabulary Squadron parades in Berlin. (*U.S. Army Signal Corps*)

Division) and five battalions of tank destroyers, the 601st, 701st, 804th TD Bn (SP) and the 805th TD Bn (Towed) used the vehicles in Italy. A single separate cavalry unit remained in Italy, the 91st Cavalry Reconnaissance Squadron (Mechanized), of the Fifth U.S. Army, which included the Brazilian Expeditionary Force as of September 1944, also equipped with American armored cars.

GENERALS AND ARMORED CARS

High-ranking U.S. Army officers frequently used M8 and M20 armored cars.

In a way, this was in response to the criteria required in 1942 with the T22 study. For certain generals, these fast armored cars were a way of getting as close as possible to the front-line troops, while taking advantage of some protection; for others it was an attribute of command, almost an idea of grandeur, in the same way the chariot was to Roman emperors.

Eupen, Belgium, 19 September 1944. Maj-Gen Joseph L. Collins, commanding VII Corps, talks with officers. On the front of his M20 is fixed a plate denoting the two stars of his rank. At the time, VII Corps numbered four infantry divisions (1st, 9th, 83rd and 104th), two armored divisions (3rd and 5th), plus the 4th Cavalry Group. (*ACME ref. 737406*)

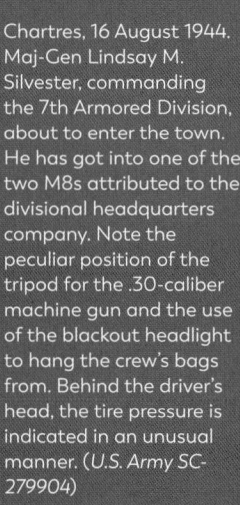

Chartres, 16 August 1944. Maj-Gen Lindsay M. Silvester, commanding the 7th Armored Division, about to enter the town. He has got into one of the two M8s attributed to the divisional headquarters company. Note the peculiar position of the tripod for the .30-caliber machine gun and the use of the blackout headlight to hang the crew's bags from. Behind the driver's head, the tire pressure is indicated in an unusual manner. (*U.S. Army SC-279904*)

Above: The Army Chief of Staff, Gen George C. Marshall, inspecting the Siegfried Line on 12 October 1944. He is on board an M20 of VII Corps, personalized by the four stars of his rank. He is accompanied by Major General Collins (in the turret facing the camera) and Lt-Gen Walter Bedell-Smith, Eisenhower's Chief of Staff. Among the peculiarities are the handle welded level with the star to make climbing aboard easier, and the double siren. (*U.S. Army SC-194294*)

Left: Lieutenant General George S. Patton made a habit of travelling in specially converted vehicles. After his Jeep, his M3A1 Scout Car or his WC-57 Dodge, it was an M20 Armored Utility Car which he used at the end of 1944 for his outings close to the front. The markings on the sides were, at bottom left, the shipping stencil (dimensions, weight, etc.), and, bottom right, the unit serial number 48623 of the Third U.S. Army Headquarters, accompanied by the relevant color coding. (*U.S. Army SC-232852*)

Above: Another view of Patton's M20 with its impressive windshield. He is accompanied by Brigadier James F. Gault (Scots Guards, center), Eisenhower's British Attaché. Note the small step to the right of the headlight protection. (*U.S. Army SC-232768*)

Right: General Patton was consistent with the image he projected of himself, his personal vehicles always being modified to a certain theme—large pennants, with the three stars of his rank on a red background, on the right side, and that of his command, the Third U.S. Army, on the left, plus long horn trumpets and sirens. (*U.S. Army SC-232767*)

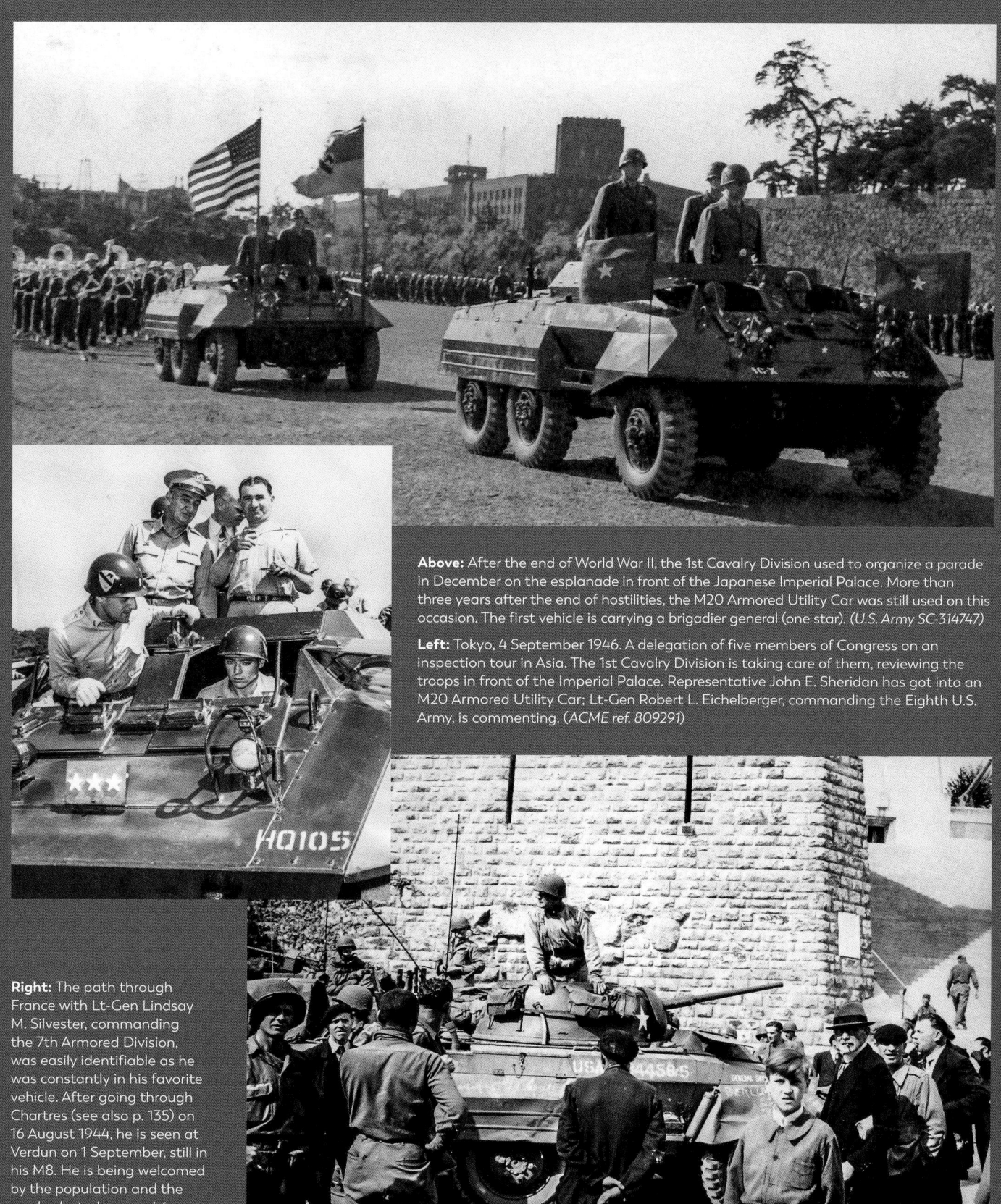

Above: After the end of World War II, the 1st Cavalry Division used to organize a parade in December on the esplanade in front of the Japanese Imperial Palace. More than three years after the end of hostilities, the M20 Armored Utility Car was still used on this occasion. The first vehicle is carrying a brigadier general (one star). (*U.S. Army SC-314747*)

Left: Tokyo, 4 September 1946. A delegation of five members of Congress on an inspection tour in Asia. The 1st Cavalry Division is taking care of them, reviewing the troops in front of the Imperial Palace. Representative John E. Sheridan has got into an M20 Armored Utility Car; Lt-Gen Robert L. Eichelberger, commanding the Eighth U.S. Army, is commenting. (*ACME ref. 809291*)

Right: The path through France with Lt-Gen Lindsay M. Silvester, commanding the 7th Armored Division, was easily identifiable as he was constantly in his favorite vehicle. After going through Chartres (see also p. 135) on 16 August 1944, he is seen at Verdun on 1 September, still in his M8. He is being welcomed by the population and the newly elected mayor, Léon Chaize. (*U.S. Army SC-279910*)

Arriving in England

The M8s and M20s arrived in England rather late, production starting in 1943 and the long supply channels preventing them from being delivered before 1944.

Above: In the fields near Ballykinler in Northern Ireland, 15 March 1944. The 3rd Battalion, 11th Infantry Regiment, 5th Infantry Division, has been assembled for a lesson in how to identify vehicles from the divisional reconnaissance troop. (*U.S. Army SC-314303*)

Right: The English population suffered enormously, and its orphans were often sponsored by American units, like the 66th Armored Regiment, 2nd Armored Division, who have done them the honor of allowing them up onto an M8 on 2 April 1944. Joyce, Patricia P., and Walter have got into the turret; David has taken the driver's place and Patricia L. the co-driver's. As this photograph was intended for the press, the tactical markings have been spattered with mud. (*U.S. Army SC-293710*)

Left: Preventive maintenance was absolutely vital for keeping the vehicles in working order; the units could not afford to break down when out on missions. On 9 March 1944, at Seaford (Northern Ireland) with elements of the 5th Reconnaissance Troop (Mechanized), 5th Infantry Division. (*U.S. Army SC-314213*)

Below: In an English village somewhere, 20 May 1944, a column moves to its assembly zone before loading up for France. Much of the country was affected by this mobilization, under the calm and serene gaze of the inhabitants. At each crossroads, a forest of signposts was planted and most of the smaller roads turned into one-way routes to keep the incessant convoy traffic flowing. (*U.S. Army SC-190408*)

Right: Gaeta, 45 miles north of Naples, 20 May 1944. These vehicles belong to an unidentified tank destroyer battalion, of which we can only see the letters TD on the front right-hand side of the vehicle. On the M8's side, note the large rack for the traditional hand tools. (*U.S. Army Signal Corps*)

The Italian Campaign

Italy was the first theater of operations in which the M8 and M20 Armored Cars were engaged, but they only arrived sparingly at the very end of 1943.

What is paradoxical is that historians generally agree they entered the war during the Battle of Monte Cassino, but without agreeing on a date, not even a month. As this battle lasted from January to May 1944, this has left a considerable number of dates to choose from.

Left: With a very appropriate nickname— "Conquistador"—this M8 from C Troop, 117th Cavalry Reconnaissance Squadron, has just entered the small village of Norma on 30 May 1944. Assembling in front of the public fountain in the village square, the population inspects the first vehicle of the liberation and its crew. (*U.S. Army SC-191120*)

Above: The advance on Rome was not easy, as this smoking, trackless M5 Light Tank on the side of highway No. 7 shows. The reconnaissance units advanced cautiously and the gaps between the vehicles were prudently lengthened on 4 May 1944. (*U.S. Army SC-191013*)

Left: The 91st Cavalry Reconnaissance Squadron cautiously reaches the outskirts of Rome on 4 June 1944. The Fifth U.S. Army under Maj-Gen Mark W. Clark had just broken through the final lines of defense before moving into the open city. (*U.S. Army SC-191060*)

Above left: On the road to Castiglioncello, to the south of Livorno in Tuscany, on 13 July 1944, a mine has got the better of this M8. While Sgt David D. Cushman inspects the damage, Pvt Raymond A. Bates remains on alert behind the 37-mm gun. (*U.S. Army SC-191013*)

Above right: The side of this vehicle has been hit; the tripod of the .30-caliber machine gun has almost been blown off. The "R36" markings designate the vehicle from a troop of an infantry division. (*U.S. Army Signal Corps*)

Right: Entering Rome from the southeast and the Via Tuscolana on 4 June 1944. The Italians are jubilant; this M8 is only a stone's throw from Mount Palatine and the most famous palace of antiquity. A bit further along, straight ahead, it will reach the Colosseum. (*U.S. Army SC-191054/055*)

Above: The Cassino sector, 20 February 1944. The M8 "Argonaught" is from the 91st Cavalry Reconnaissance Squadron, a unit of the Fifth Army. (*NARA*)

Right: The 81st Cavalry Reconnaissance Squadron (1st Armored Division) advances along an Italian road just before the capture of Rome, leaving a broken-down jeep in its wake. (*U.S. Army Signal Corps*)

The French Campaign

Right: Mid-July 1944. After very hard fighting, the 29th Infantry Division finishes off the capture of Saint-Lô. The vehicles of the 29th Reconnaissance Troop gather among the ruins of the town. (*Signal Corps*)

Below: Troop B of the 4th Cavalry Reconnaissance Squadron. With the 24th Cavalry Reconnaissance Squadron, it formed the 4th Cavalry Group, Mechanized (First U.S. Army under General Hodges). The unit serial number 44883 on the front right of the vehicle confirms the unit. (*U.S. Army Signal Corps*)

Above: This M8 belongs to Troop A of the 106th Cavalry Reconnaissance Squadron which formed the 106th Cavalry Group (the markings on the bottom of the forward hull) with the 121st Cavalry Reconnaissance Squadron. The 106th was a unit from XV Corps (Seventh Army, General Patch) as is demonstrated by the "US XV" in a rather unconventional position between the two drivers. (*U.S. Army Signal Corps*)

Left: August 1944. Gen Lindsay McD. Silvester, the 7th Armored Division's Commander, enters freshly liberated Chartres aboard his command M8. (*Signal Corps*)

Right: Elements of the 2nd Armored Division and the 4th Infantry Division liberate the little Norman village of Canisy on 29 July 1944, 5.5 miles to the southeast of Saint-Lô. This M8 is part of C Platoon, 82nd Armored Reconnaissance Battalion; the markings on the side are typical of this unit. (*U.S. Army SC-192412*)

Below: A jubilant population with presents for the liberators were familiar scenes oft repeated like here in the outskirts of Bréhal, 2 August 1944, with this Norman lady offering cider to the men of the 42nd Cavalry Reconnaissance Squadron. With the 2nd Cavalry Reconnaissance Squadron, it formed the 2nd Cavalry Group of XII Corps of the Third U.S. Army (Patton). (*U.S. Army SC-192253*)

Above: Mid-June 1944. An M20 from the 801st Tank Destroyer Battalion (First Army) enters Montebourg, in the Manche Department. (*Signal Corps*)

Left: La Haye du Puits, in the Cotentin Peninsula, 10 July 1944. Men from the 79th Infantry Division examine the effects of a mine on an M8 Light Armored Car most likely belonging to the 79th Cavalry Reconnaissance Troop. Despite the strength of the explosion, the vehicle seems to have maintained its shape and the armor plates are not dislocated. The fire that followed did more damage. (*U.S. Army SC-191375*)

Above: The fighting has been very hard for these M8s assembled near Metz. There are more than one hundred waiting to be transferred or repaired. They are a mix of early and late production vehicles. Some bear the windshield kit and others have the rarely seen D67511 Folding Pintle Bracket on the turret. (*U.S. Army Signal Corps*)

Right: The reconnaissance units did not always range as far as was expected "by the book"; they sometimes advanced very cautiously as their armament was, above all, defensive. (*U.S. Army Signal Corps*)

Above: In the shelter of Block No. 2 of the Maginot Line at Rohrbach (Fort Casso), 13 December 1944. The 92nd Cavalry Reconnaissance Squadron of the 12th Armored has regrouped for servicing and re-supply. The enemy is still very close as a guard is keeping watch from the top of a hillock. (*U.S. Army SC-197386*)

Left: In the little village of Xousse, near Lunéville, 18 November 1944. Elements of the 63rd Engineer Combat Battalion of the 44th Infantry Division pass through the completely destroyed village. This M8 from an unidentified unit, which is overtaking the convoy, has lost all its mudguards. (*U.S. Army SC-196654*)

Winter 1944–45

Right: This deer hunted in the deep Ardennes forests is enough to improve daily rations. The lack of space in the armored cars made a lot of crews create makeshift baggage racks on the rear decks of their vehicle, this one being ideal to transport the animal. (*U.S. Army Signal Corps*)

Below: In the hamlet of Borzée, near La Roche-en-Ardenne, Luxemburg, 16 January 1945. An M8 of the 4th Cavalry Reconnaissance Squadron has a re-supply and weapons-check break. To provide more storage space, a metal box has been welded onto the back of the turret. (*U.S. Army Signal Corps*)

Left: Saarlauten, Germany, 13 January 1945. Sergeants C. B. Hennemen and J. C. Espy apply winter camouflage to the bodywork of one of the three M20s of the Headquarters Company of the 607th Tank Destroyer Battalion. Between 11 November 1944 and 2 February 1945, the 607th TD was attached to the 95th Infantry Division. (*U.S. Army Signal Corps*)

Below: Two M8s from the 2nd Cavalry Reconnaissance Squadron patrolling in the sector held by the 4th Infantry Division on 14 January 1945. Associated with the 42nd Cavalry Reconnaissance Squadron, it formed the 2nd Cavalry Group. Note how carefully the white paint has been applied without obscuring the unit markings or stars. (*U.S. Army Signal Corps*)

Right: A wintry meeting along the River Ourthe in Belgium, 16 January 1945. Infantrymen Rodney Himes and Alfred Gernhardt from the 84th Infantry Division cordially greet T/5 Ancel Carey aboard his armored car from the 41st Cavalry Reconnaissance Squadron (11th Armored Division). This handshake marks the meeting of two army corps, VII Corps for the infantry and XII Corps for the cavalry. (*U.S. Army SC-199155*)

Below: Germany, 26 January 1945. "No stopping" says the German signpost, which visibly amuses this M8's crew. While Cpl Don MacClure gets down from the vehicle under the watchful eye of Lt George F. Rork in the turret, Corporals Benny Klucinec and Lohman Bacly in the drivers' seats watch the vicinity. The three soldiers appear to be wearing British motorcyclist helmets. (*U.S. Army SC-194443*)

Above: January 1945 near Michelau in the region of Diekirch, Luxemburg. A reconnaissance column from the Third U.S. Army comes to a stop. All the M8s have been given a rear baggage rack and winter camouflage. (*U.S. Army Signal Corps*)

Left: In the Wittlich region, Germany, 10 March 1945. Winter is coming to an end and troop movements are easier. The 90th Cavalry Reconnaissance Troop moves through a devastated forest. (*U.S. Army SC-201951*)

Below: This M20 Armored Utility Car, seen on 17 February 1945, belongs to the 6th Cavalry Reconnaissance Group of the Third Army. Note that a makeshift transparent shield has been placed in front of the combat compartment, as has a step under the forward hull. (*U.S. Army Signal Corps*)

Germany: the Final Round

Right: Watched by a stunned population, this column of Company A, 82nd Armored Reconnaissance Battalion, penetrates a town where each house has hung out a white sheet as a sign of surrender. From the position of the gun sight for the cannon, this M8 is an early production vehicle. (*U.S. Army Signal Corps*)

Below: Neustadt, near Mannheim, Germany, 22 March 1945. While waiting for the advance to start again, men of the 328th Infantry Regiment of the 26th Infantry Division take a break while the mortars free the terrain in front of them. (*U.S. Army SC-270912*)

Left: Germany, March 1945. An M8 of the 94th Cavalry Reconnaissance Squadron unloads a prisoner. At the time, the unit was assigned to Combat Command A of the 14th Armored Division. (*Signal Corps*)

Below: Setterich to the north of Aachen (Aix-la-Chapelle). M8s advance cautiously among the ruins. The crews are tense; the armored cars were not designed for this type of mission; their armor is too light to counter anti-tank weapons and they do not have enough room to maneuver. They can only count on their speed to get out of an ambush. (*U.S. Army SC-271912*)

Above: Troop C of the 24th Cavalry Reconnaissance Squadron, First U.S. Army, in the village of Grevenstroich. On returning from a patrol, the crew of this M8 has brought back two prisoners who are disembarked near the platoon's half-track. (*U.S. Army SC-380397*)

Right: A cautious entry into the German village of Ehrang, near Trier. The crew have spread the fluorescent cherry red air-to-ground identification flag over the engine compartment. (*U.S. Army Signal Corps*)

Above: Kessler, Germany, 12 April 1945. Reconnaissance vehicles, not front-line combat vehicles; as soon as there is a confrontation, the armored cars move back, and the tanks move up. (*U.S. Army SC-380397*)

Below: Austria, May 1945. Accompanied by the 614th and 781st Tank Destroyer battalions, the 103rd Cavalry Reconnaissance Troop penetrates into Scharnitz. Their mission was to reach the banks of the Danube as quickly as possible. (*U.S. Army SC-380397*)

Right: Ahlen, Germany, 1 April 1945. The Ninth U.S. Army enters the town. Despite the white flags and deserted roads, the tension is palpable. (*U.S. Army SC-332971*)

Below: A reconnaissance unit takes possession of the freight station at Rheinberg, near Düsseldorf in the north of the Rhineland, 17 March 1945. It has been a long road to get there; the men use the station's water system for refilling steam locomotives to wash their vehicles. (*U.S. Army SC-255938*)

The Spoils of War

Using materiel recovered from the enemy was common in all armies and the M8 Light Armored Cars were no exception.

Several examples were recovered by German forces and turned against their former owners. At least one example underwent extensive modifications.

On 17 July 1945, Lt J. F. Eppes and PFC W. F. Wilcock, Ordnance Technical Intelligence Team No. 15, after a quick study, produced Ordnance Technical Intelligence Report No. 351 on an M8 modified by the German Army.

They noted all the armament had been removed and that the front of the turret had been covered with a 1.5-inch-thick steel plate. The original plate which covered part of the roof of the turret had been removed as had the swiveling mechanism, which meant it was free turning. Halfway down inside the turret, a new floor had been installed to take a triple 20-mm gun carriage, the 20-mm MG151/20 *Flakdrilling*, which was normally installed in the SdKfz 251/21.

Extracted from Ordnance Technical Intelligence Report No. 351, the three photographs that accompanied the report. There was only one 20-mm gun on the carriage. (*Ordnance*)

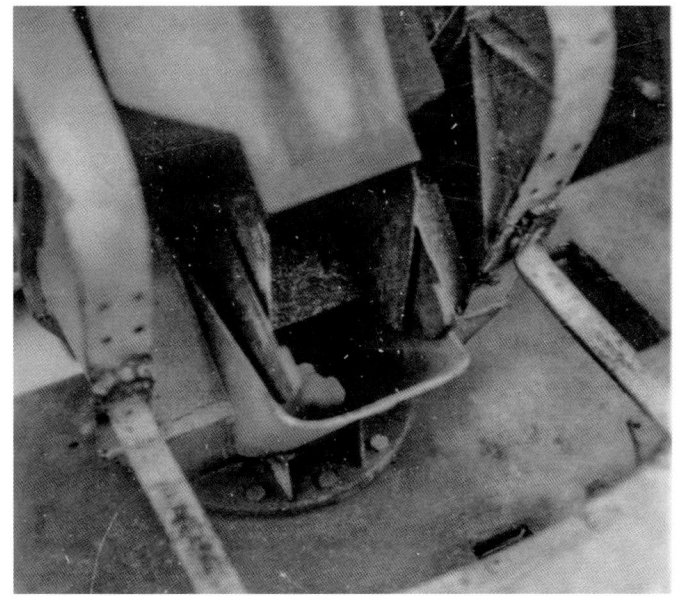

Above: The German Army started reusing the M8 practically as soon as they appeared in Northwest Europe. This shot was taken in Lorraine during the summer of 1944 and shows a vehicle from the 42nd Cavalry Reconnaissance Squadron. (*German photo*)

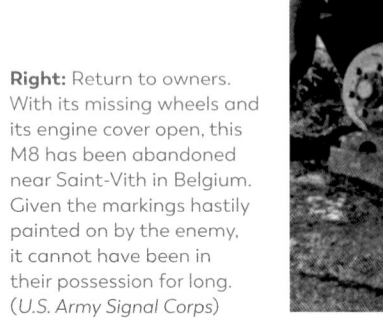

Right: Return to owners. With its missing wheels and its engine cover open, this M8 has been abandoned near Saint-Vith in Belgium. Given the markings hastily painted on by the enemy, it cannot have been in their possession for long. (*U.S. Army Signal Corps*)

Above: The last major attack by German forces, in the Ardennes, was an opportunity to capture a significant number of armored cars. This shot, taken from a famous documentary, shows a group of *Waffen-SS* in front of an apparently intact M8. It is 18 December 1944 at Poteau where a column of the 14th Cavalry Group has been destroyed. (*U.S. Army SC-198249*)

Left: Faymonville, Belgium, 18 January 1945. T/5 James W. King of the 1st Engineer Combat Battalion, 1st Infantry Division, examines an M8 which his division has just neutralized. (*U.S. Army SC-199163*)

Middle and right: These were fortuitous prizes and saw opportunistic use on the part of the Germans. There was no intention of forming any sort of group with these captured vehicles, except for those with Skorzeny in the Ardennes. They were not modified in any way and even kept the equipment that was attached to them. (*DR*)

After 1945

Below: June 1945, the 3rd Armored Regiment sets itself up in Germany to form part of the occupying force. (*Private Collection*)

Above: The occupation of Germany started after the end of hostilities; keeping order was ensured by the mobile lightly armored units. Sergeants Wallace, Hitzler, Elliot and Mason take possession of a brand-new Ford M8 as detailed by Major Delcorso. This example is a late production vehicle with its locker for the drivers' windshields and the siren replaced by the horn. (*Cleveland Press*)

Below left: It was not until July 1946 that the regular occupation units were replaced in their policing role by the U.S. Constabulary, whose colors (yellow and blue) were painted in very visible stripes on this M8 of the 13th Squadron. (*Private Collection*)

Below right: It was still a complicated process to protect an M20 from heavy rain. This example assigned to the military police comes from the 773rd Tank Destroyer Battalion, stationed in Berlin. (*Private Collection*)

Left: Germany, June 1945. This M20 belongs to the 94th Cavalry Reconnaissance Squadron, 14th Armored Division. The crew has solved the problem of the lack of space by installing a structure above the engine covers. A step as well as a handle has been welded to the middle of the star to make getting into the vehicle easier. (*Private Collection*)

Below: Berlin, 26 June 1948. A patrol from the Constabulary Force at the foot of the Tiergarten Monument to the glory of the Red Army. (*NY Times ref. L130.MCM*)

Above: Miles away from Europe, the M8s and M20s also formed the backbone of the forces keeping order in Asia. This M8 from the 24th Cavalry Reconnaissance Troop is patrolling in Kyushu, Japan, in April 1947. (*Private Collection*)

Right: A highly tense surveillance of a demonstration for the victims of fascism in the Russian sector of Berlin, 12 September 1948. The front of the old Berlin Museum was covered with the flags of 20 nations; strangely enough, that of the USA was absent, according to the reporters from the *New York Times* Paris office. The M8 bears the markings of Berlin Command. (*NY Times ref. 40.MA*)

Appendix

#	Deliveries from	Deliveries to	Transfer Date	Remarks and comments
	DELIVERIES AND DESTINATIONS OF THE 159 TEST VEHICLES			
#6	Saint Paul	APG Aberdeen	8 April 1943	Transferred to Ford Highland Park for 10 days then replaced by the #46.
#7	Chicago	LTD Lima	26 May 1943	Engineering test + Tires 11.00×18 test + 5,000 miles durability test.
#8	Saint Paul	APG Aberdeen	8 April 1943	Firing test.
#9	Chicago	LTD Lima	26 May 1943	
#10	Saint Paul	GMPG Milford	8 April 1943	(USA 6032234) Engineering test.
#11	Chicago	LTD Lima	25 May 1943	Training Film Project at Camp Young.
#12	Saint Paul	LTD Lima	9 April 1943	Transferred to the Armored Force Board at Fort Knox, Kentucky. SCR-510 tests. General vehicle overhaul at the Chester Tank Depot.
#13	Chicago	LTD Lima	25 May 1943	Desert Warfare Board at Camp Young (April 1944).
#14	Saint Paul	LTD Lima	9 April 1943	Transferred to the Armored Force Board at Fort Knox, Kentucky. SCR-510 tests. General vehicle overhaul at the Chester Tank Depot.
#15	Chicago	GMPG Milford	24 May 1943	
#16	Saint Paul	LTD Lima	9 April 1943	Transferred to the Armored Force Board at Fort Knox, Kentucky. SCR-510 Tests. General vehicle overhaul at the Chester Tank Depot.
#17	Chicago	TAPG Utica	24 May 1943	Tests 10,000 miles.
#18	Saint Paul	LTD Lima	9 April 1943	Transferred to the Armored Force Board at Fort Knox, Kentucky. SCR-528 tests. General vehicle overhaul at the Chester Tank Depot.
#19	Chicago	LTD Lima	25 May 1943	Desert Warfare Board at Camp Young (April 1944).
#20	Saint Paul	LTD Lima	9 April 1943	Transferred to the Armored Force Board at Fort Knox, Kentucky. SCR-528 tests. General vehicle overhaul at the Chester Tank Depot.
#21	Chicago	LTD Lima	25 May 1943	Desert Warfare Board at Camp Young (April 1944).
#22	Saint Paul	LTD Lima	9 April 1943	Transferred to the Armored Force Board at Fort Knox, Kentucky. SCR-528 tests. General vehicle overhaul at the Chester Tank Depot.
#23	Chicago	LTD Lima	25 May 1943	
#24	Saint Paul	LTD Lima	10 April 1943	Transferred to Cavalry Board at Fort Riley, Kansas. SCR-506 tests.
#25	Chicago	LTD Lima	25 May 1943	
#26	Saint Paul	LTD Lima	10 April 1943	Transferred to Cavalry Board at Fort Riley, Kansas. SCR-506 tests.
#27	Chicago			Not transferred. Hercules diesel JXLD engine trials
#28	Saint Paul	LTD Lima	12 April 1943	(USA 3032252) Transferred to Tank Destroyer Board at Camp Hood, Texas. SCR-608 tests.

#29	Chicago	LTD Lima	25 May 1943	(USA-6032253) Engine change.
#30	Saint Paul	LTD Lima	12 April 1943	Transferred to Tank Destroyer Board to the Camp Hood, Texas. SCR-610 tests.
#31	Chicago	LTD Lima	25 May 1943	
#32	Saint Paul	Rouge Plant	12 April 1943	500-mile test for oil and petrol consumption.
#33	Chicago	LTD Lima	26 May 1943	Winterization Project.
#34	Saint Paul	Rouge Plant	12 April 1943	Engineering test.
#35	Chicago	LTD Lima	26 May 1943	Defective turret repaired at the Chester Tank Depot.
#36	Saint Paul	CTD Chester	16 April 1943	
#37	Chicago	LTD Lima	26 May 1943	General vehicle overhaul at the Chester Tank Depot.
#38	Saint Paul	CTD Chester	16 April 1943	
#39	Chicago	LTD Lima	26 May 1943	
#40	Saint Paul	CTD Chester	14 April 1943	
#41	Chicago	LTD Lima	26 May 1943	
#42	Saint Paul	APG Aberdeen	19 April 1943	
#43	Chicago	LTD Lima	26 May 1943	Defective turret repaired at the Chester Tank Depot.
#44	Saint Paul	APG Aberdeen	19 April 1943	(USA 6032268) test Ring Mount M49C.
#45	Chicago	LTD Lima	26 May 1943	Defective turret repaired at the Chester Tank Depot.
#46	Saint Paul	TAPG Utica	23 April 1943	Transferred to Ford Highland Park to replace #6. 5,000-mile durability test + replacing front suspension, 10,000-mile tests.
#47	Chicago	LTD Lima	26 May 1943	(USA 6032271) Return to factory, 15 September for serious turret damage.
#48	Saint Paul	CTD Chester	23 April 1943	
#49	Chicago	LTD Lima	26 May 1943	Defective turret repaired at the Chester Tank Depot.
#50	Saint Paul	LTD Lima	21 April 1943	Training Film Project.
#51	Chicago	LTD Lima	26 May 1943	
#52	Saint Paul	CTD Chester	23 April 1943	Defective turret repaired in the field.
#53	Chicago	LTD Lima	26 May 1943	Defective turret repaired at the Chester Tank Depot.
#54	Saint Paul	CTD Chester	24 April 1943	
#55	Chicago	LTD Lima	26 May 1943	
#56	Saint Paul	CTD Chester	23 April 1943	
#57	Chicago	LTD Lima	26 May 1943	Defective turret repaired at the Chester Tank Depot.
#58	Saint Paul	CTD Chester	26 April 1943	Defective turret repaired in the field.
#59	Chicago	LTD Lima	26 May 1943	Defective turret repaired at the Chester Tank Depot.
#60	Saint Paul	CTD Chester	24 April 1943	Defective turret repaired in the field.
#61	Chicago	LTD Lima	26 May 1943	Defective turret repaired at the Chester Tank Depot.
#62	Saint Paul	CTD Chester	24 April 1943	
#63	Chicago	LTD Lima	26 May 1943	
#64	Saint Paul	CTD Chester	23 April 1943	
#65	Chicago	LTD Lima	26 May 1943	
#66	Saint Paul	CTD Chester	26 April 1943	
#67	Chicago	LTD Lima	26 May 1943	Defective turret repaired at the Chester Tank Depot.
#68	Saint Paul	CTD Chester	24 April 1943	

#69	Chicago	LTD Lima	26 May 1943	Defective turret repaired at the Chester Tank Depot.
#70	Saint Paul	CTD Chester	11 May 1943	Defective turret repaired in the field.
#71	Chicago	LTD Lima	26 May 1943	
#72	Saint Paul	CTD Chester	27 April 1943	
#73	Chicago	LTD Lima	27 May 1943	Defective turret repaired at the Chester Tank Depot.
#74	Saint Paul	CTD Chester	27 April 1943	
#75	Chicago	LTD Lima	27 May 1943	
#76	Saint Paul	CTD Chester	27 April 1943	
#77	Chicago	LTD Lima	27 May 1943	Defective turret repaired at the Chester Tank Depot.
#78	Saint Paul	CTD Chester	29 April 1943	
#79	Chicago	LTD Lima	27 May 1943	Defective turret repaired at the Chester Tank Depot.
#80	Saint Paul	CTD Chester	27 April 1943	
#81	Chicago	LTD Lima	27 May 1943	Defective turret repaired at the Chester Tank Depot.
#82	Saint Paul	CTD Chester	29 April 1943	
#83	Chicago	LTD Lima	27 May 1943	Defective turret repaired at the Chester Tank Depot.
#84	Saint Paul	CTD Chester	29 April 1943	
#85	Chicago	LTD Lima	27 May 1943	Defective turret repaired at the Chester Tank Depot.
#86	Saint Paul	CTD Chester	14 April 1943	Tests 20,000 miles.
#87	Chicago	LTD Lima	27 May 1943	Defective turret repaired at the Chester Tank Depot.
#88	Saint Paul	CTD Chester	29 April 1943	(USA 6032312) sent to the 5th Infantry Division for training.
#89	Chicago	LTD Lima	27 May 1943	
#90	Saint Paul	CTD Chester	30 April 1943	
#91	Chicago	LTD Lima	27 May 1943	Defective turret repaired at the Chester Tank Depot.
#92	Saint Paul	CTD Chester	30 April 1943	
#93	Chicago	LTD Lima	27 May 1943	Defective turret repaired at the Chester Tank Depot.
#94	Saint Paul	CTD Chester	11 May 1943	Defective turret repaired in the field.
#95	Chicago	LTD Lima	27 May 1943	Defective turret repaired at the Chester Tank Depot.
#96	Saint Paul	CTD Chester	12 May 1943	Defective turret repaired in the field.
#97	Chicago	LTD Lima	2 June 1943	
#98	Saint Paul	CTD Chester	12 May 1943	Defective turret repaired in the field.
#99	Chicago	LTD Lima	2 June 1943	
#100	Saint Paul	CTD Chester	12 May 1943	Defective turret repaired in the field.
#101	Chicago	LTD Lima	2 June 1943	
#102	Saint Paul	CTD Chester	12 May 1943	Defective turret repaired in the field.
#103	Chicago	LTD Lima	2 June 1943	
#104	Saint Paul	CTD Chester	15 May 1943	Defective turret repaired in the field.
#105	Chicago	LTD Lima	2 June 1943	
#106	Saint Paul	CTD Chester	12 May 1943	Defective turret repaired in the field.
#107	Chicago	LTD Lima	2 June 1943	
#108	Saint Paul	CTD Chester	12 May 1943	Defective turret repaired in the field.
#109	Chicago	LTD Lima	2 June 1943	

#110/112/114	Saint Paul	CTD Chester	15 May 1943	Defective turret repaired in the field.
#116/118/120	Saint Paul	CTD Chester	15 May 1943	Defective turret repaired in the field.
#122	Saint Paul	CTD Chester	21 May 1943	Defective turret repaired in the field.
#124/126/128	Saint Paul	CTD Chester	15 May 1943	Defective turret repaired in the field.
#130/132/134	Saint Paul	CTD Chester	15 May 1943	Defective turret repaired in the field.
#136	Saint Paul	CTD Chester	21 May 1943	
#138/140/142	Saint Paul	CTD Chester	15 May 1943	Defective turret repaired in the field.
#144	Saint Paul	CTD Chester	15 May 1943	Defective turret repaired in the field.
#146	Saint Paul	CTD Chester	21 May 1943	Defective turret repaired in the field.
#148/150/152	Saint Paul	CTD Chester	15 May 1943	Defective turret repaired in the field.
#154	Saint Paul	CTD Chester	21 May 1943	Defective turret repaired in the field.
#156/158	Saint Paul	CTD Chester	15 May 1943	Defective turret repaired in the field.
#160/162	Saint Paul	CTD Chester	21 May 1943	Defective turret repaired in the field.
#164/166	Saint Paul	CTD Chester	14 May 1943	Defective turret repaired in the field.
#168/170/172	Saint Paul	CTD Chester	21 May 1943	Defective turret repaired in the field.
#174/176/178	Saint Paul	CTD Chester	21 May 1943	Defective turret repaired in the field.
#180/182/184	Saint Paul	CTD Chester	21 May 1943	Defective turret repaired in the field.
#186	Saint Paul	CTD Chester	21 May 1943	Defective turret repaired in the field.
#188/190/192	Saint Paul	CTD Chester	22 May 1943	Defective turret repaired in the field.
#194/196/198	Saint Paul	CTD Chester	22 May 1943	Defective turret repaired in the field.
#200	Saint Paul	CTD Chester	22 May 1943	
#202/204/206	Saint Paul	CTD Chester	22 May 1943	Defective turret repaired in the field.
#210/212/214	Saint Paul	CTD Chester	22 May 1943	Defective turret repaired in the field.
#216/218	Saint Paul	CTD Chester	22 May 1943	Defective turret repaired in the field.
#220/222/224	Saint Paul	CTD Chester	22 May 1943	Defective turret repaired in the field.

Bibliography

Main Technical Handbooks

FM 2-6 Crew Drill, M8 Light Armored Car, December 1943.

FM 2-20 Cavalry Reconnaissance Troop Mechanized, 24 February 1944.

FM 2-30 Cavalry Mechanized Reconnaissance Squadron, 29 March 1943.

SNL G-136 M8 Light Armored Car, 1 August 1945.

SNL G-136 M8 Light Armored Car, 12 December 1951.

SNL G-176 M20 Armored Utility Car, 1 August 1945.

SNL G-176 M20 Armored Utility Car, 12 December 1951.

TM 9-743 (French Army) M8 Light Armored Car and M20 Armored Vehicle all uses, 21 February 1944.

TM 9-743 M8 Light Armored Car, 10 March 1943.

TM 9-1743 Power Train, Suspension, Hull, and Turret for M8 Light Armored Car and M20 Armored Utility Car, 26 October 1943.

TM 9-1904 Ammunition Inspection Guide, 2 March 1944.

TM 9-2700 Principles of Automotive Vehicles, 18 November 1947.

TM 9-2800 Military Vehicles, October 1947.

TM 9-2800 Standard Military Motor vehicles, 1 September 1943.

TM 9-2800-1 Military Vehicles, September 1953.

TM 9-2853 Preparation of Ordnance Materiel for Deep Water Fording, 7 July 1945.

TM 11-227 Radio Communication Equipment, 10 April 1944.

TM 11-702 Interphone Equipment, RC-99, 3 May 1944.

Historical Documentation

Aberdeen Combat Vehicle Test Board, final evaluation report on the T22E2 pre-production series vehicle, 27 November 1942.

Aberdeen Proving Ground, report on the daily development of the test on the preproduction series T22E2, December 1942.

Army Ground Forces Board, evaluation report on the T22E2 pre-production series vehicle, 24 May 1942.

Army Ground Forces Board, evaluation report on the T22E2 pre-production series vehicle, 8 September 1942.

Design, Development, Engineering and Production of Armored Cars 1940-1944, 28 October 1944.

Development of Armored Vehicles, Armored Car, Scout Car & Personnel Carrier, 12 September 1949.

Office of the Chief of Ordnance, memorandum, and final evaluation report on the T22 prototype, 25 March 1942.

Office of the Chief of Ordnance, report on the internal layout of the production series M8s, 30 November 1942.

Reports, evaluations, and exchanges of letters between the various components of the Army, Ordnance and civilian companies, gathered together by First Lieutenant J. R. Murray, Wheeled Vehicles Section, Engineering Branch of Ordnance, 1941–45.

Tank Destroyer Command, evaluation report prototype, 15 May 1942.

Austria, September 1948. The 24th Constabulary Squadron leaves its quarters to take part in an exercise. (*Signal Corps*)

Reference Books

Crismon, Fred W. *US Military Wheeled Vehicles*. Osceola, WI: Motorbooks International, 1994.

Engineering of Transport Vehicles, 1942–1945 (Chief of Ordnance—Detroit), 1945.

Lend-Lease Shipments World War II (Office Chief of Finance, War Department), 31 December 1946.

Official Production of the United States, July 1, 1940 to August 31, 1945, 1 May 1947.

Ordnance Department, Administrative and Tactical Vehicles, 1940–1944 (Automotive Center), 1 January 1944.

Ordnance Department, Administrative and Tactical Armored Vehicles, 1940–1945 (Automotive Center), 1 May 1945.

Ordnance Department, Administrative and Tactical Vehicles, 1940–1945 (Automotive Center), 1 October 1945.

Summary Report of Acceptances, Tank-Automotive Material, 1940–1945 (Chief of Ordnance—Detroit), 1945.

Vanderveen, Bart. *Historic Military Vehicles Directory*. London: Battle of Britain Prints International, 1989.

Vanderveen, Bart. *The Observer's Fighting Vehicles Directory*. London: Frederick Warne & Co Ltd, 1969.